THE CAR SECURITY MANUAL

HOW TO PROTECT YOUR CAR FROM THIEVES & JOYRIDERS

David Pollard

VELOCE PUBLISHING PLC

PUBLISHERS OF FINE AUTOMOTIVE BOOKS

Also available from Veloce:

Alfa Romeo 6C by Angela Cherrett
Alfa Romeo 8C 2300 by Angela Cherrett
Alfa Romeo Giulia GT & GTA Coupé by John Tipler
British Car Factories from 1896 by Michael Stratton & Paul Collins
Bugatti 57 - The Last French Bugatti by Barrie Price
Fiat & Abarth 124 Spider & Coupé by John Tipler
The Prince & I - My Life With The Motor Racing Prince of Siam (Bira)
by Princess Ceril Birabongse
Total Tuning for the Austin Healey Sprite/MG Midget
by Daniel Stapleton

First published in 1993 by -
Veloce Publishing Plc,
Godmanstone,
Dorset DT2 7AE,
England.

ISBN 1 874105 18 9
© David Pollard & Veloce Publishing Plc 1993

Readers with ideas for automotive books, or books on other
transport or related hobby subjects, are invited to write to the
Editorial Director of Veloce Publishing at the above address.

British Library Cataloguing In Publication Data -
A catalogue record for this book is available from the British
Library.

Typesetting (in Avant Garde), design and page make-up all by
Veloce on Apple Mac.

Printed and bound in the UK.

CONTENTS
THE ROUTE TO CAR SECURITY-

PREFACE
ACKNOWLEDGEMENTS & NOTES

❏ ACKNOWLEDGEMENTS

My thanks go to the following companies for their help in compiling this book:

AAA Security, Alpine Electronics of UK Ltd., Autolok Security Products Ltd., Automaxi Ltd., Black & Decker Ltd., Blaupunkt (Robert Bosch Ltd.), Carflow Products Ltd., *Car Stereo & Security Magazine*, Clarion Shoji (UK) Ltd., Clifford Electronics Inc. (UK), Cobra (Ital Audio Ltd.), Digicom Communications, E.E.C. Ltd., Electrosystems Ltd., Fanfare Electronics, Kemitron Automotive Ltd., Klamp-It Innovations Ltd., Krooklok Ltd., Laserline Car Alarms, Liftsonic, Metro Products Ltd., Path Group Ltd., PCA Ltd., Philips Ltd., Piranha Security, Portland Marketing, Quickfit 70 Ltd., Redsure, Retainacar, Richbrook International Ltd., Ring Automotive Products Ltd., Roadstar UK Ltd., Scorpion Vehicle Security Systems, Securon Ltd., Vauxhall Motors Ltd., Westgate Motors, Northampton.

Special thanks to Ann, who did so much of the compilation and checking, and to Phil, Katie and Stuart, so willingly (?) co-opted into some very cold photographic sessions.

❏ IMPORTANT NOTE

Whilst all of the products and methods described in this book have been included in good faith, neither the publisher nor author can guarantee that your vehicle will not be attacked and/or stolen. At the time of writing all products shown are believed to be as effective as claimed and legal. However, the law, with regard to the specification and fitting of security devices, is subject to change. Establish from the vendor that any security equipment you buy conforms with current legislation.

❏ THE 'THATCHAM' REPORT

In January 1993 the Association of British Insurers (ABI) published the *British Insurance Industry's Criteria for Vehicle Security*, a complex report produced by the Motor Insurance Repair Research Centre at Thatcham.

The report evaluates current original fitment security devices and proprietory devices and marks features according to their effectivness. However, as testing and evaluation is a slow process, it is likely to be many years before the public are given the kind of information they can use to make buying choices.

The report recommends that security devices (we're really just talking about electronic systems and immobilisers) should be professionally fitted to qualify the car's owner for an insurance discount ranging between 5 and 25 per cent. It seems unlikely, however, that such a rule will be enforced when it is clearly in the insurance industry's interest for owners to improve their vehicle's security by any means.

WHY <u>YOU</u> SHOULD PROTECT YOUR CAR

SOME FACTS & FIGURES ON CAR CRIME

The massive growth in car crime throughout the 1980s has carried on relentlessly into this decade. Naturally enough, it has been followed by a similar growth in the vehicle security products industry. Though this should be good news for motorists wanting to ensure that their cars and contents stay put, the sheer number of security devices on the market has created more confusion than anything else.

It's easy to end up in a severe state of shock just browsing the car accessory shelves and, unfortunately, many owners give up in frustration and simply trust to luck that the thief will attack someone else's car. Either that or they buy an unsuitable product.

There are some excellent products around in today's marketplace; microprocessor controlled alarms are becoming the norm, with intelligent sensing and much-improved reliability. But there is a flip-side to this coin; there are plenty of systems and products out there for which the description 'over-priced junk' is a compliment! There is no excuse for buying it, because you have the choice.

As a consumer, your best defence against buying rubbish, expensive or otherwise, is to check it out *before* you buy - learning from someone else's mistakes is a darn sight less painful than learning from your own.

❏ THE BAD NEWS

The plain but unpalatable truth is simple: if a thief really wants *your* car, he'll take it - the professionals can have your pride and joy loaded on to a truck and a hundred miles away before you know it's gone.

❏ THE GOOD NEWS

You'll have to have something really

This humble Volkswagen will shop no more after being crashed by under-age 'joy-riders'.

special to warrant a thief taking that much trouble and, as most of us don't own Ferraris, the odds can be tipped heavily in your favour. By using common sense and good security products you can make someone else's car appear far more attractive to a thief than yours.

❏ THE DETERRENT

Your first line of defence is to deter the thief from even trying to steal your car and its contents. There's little point in fitting the best immobiliser in the world if the thief has to break in (doing £500 worth of damage in the process) to find it's fitted! Physical security devices should be large and brightly coloured, announcing their presence to the would-be intruder. Most electronic systems come with a flashing LED light, which should be in plain view as the thief looks through the driver's window. Alarm systems usually come with window stickers, alerting the thief to their presence. A sticker showing the manufacturer's name can have an adverse affect where a professional thief is involved - it tells him exactly what to expect. However, these stickers have the required effect on the amateur thief: he knows that taking the car will involve far more than just breaking a window. Which brings us to the second line of defence ...

❏ EASY TARGETS

If the car thief liked hard work, he wouldn't be stealing cars for a living. Likewise, getting caught and locked up is unlikely to appeal either, so he's looking for easy targets. Cars with windows left open and/or doors unlocked; cars with no security device; cars parked in dark, secluded alleys.

If this sounds like your car, start worrying now.

What he doesn't want is a siren going off loudly, lights flashing and a crowd of onlookers asking about his business!

If he thinks that anything of this nature awaits him as his reward for breaking into your car, then he's likely to pass on to the next one. After all, there's plenty of choice when it comes to unsecured vehicles. If he *does* get inside your car, he should find that you've removed your radio/cassette deck and other valuables, that hot-wiring the engine requires a degree in electronics (and an hour's free time) because of the immobilisation system in operation and basically come to the conclusion that there have to be easier cars than yours to attack. It's up to you to make sure that taking your car or contents is hard and, importantly, time-consuming work indeed.

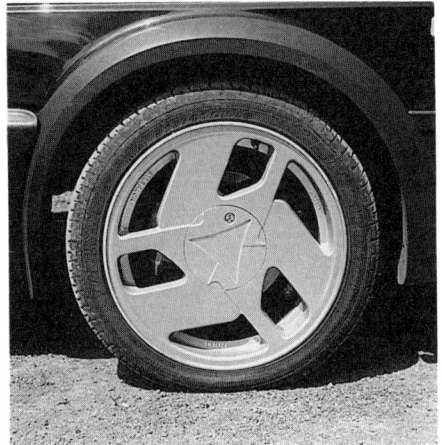

Nice alloy wheels will draw a thief like a magnet. These Nothelle wheels have a lockable centre cap, but a wise owner would fit locking wheel bolts/nuts as well.

❏ STATISTICS

As a rule, statistics of any kind make a good substitute for sleeping pills. But take a few minutes to run through these figures, taken from a UK government survey. They show that car crime is a far greater problem than is generally imagined and that the days when we could assume that it was someone else's car that would be stolen are long gone.

20% OF THEFTS TAKE PLACE IN PUBLIC CAR PARKS

What a huge selection of (someone else's) cars to choose from. It's so easy to leave something valuable on display or even leave the car unlocked; the thief relies on your inattention as you're distracted by the time, the shopping, the kids, the dentist's appointment etc.

CAR CRIME ACCOUNTS FOR ALMOST A THIRD OF ALL RECORDED CRIME IN THE UK

Incredible, but true. That's theft of cars and theft *from* cars. Your pride and joy, which cost you so much, is high on someone else's shopping list - except that they don't want to pay for it. It's up to you to make sure they don't get a free gift.

IN A THIRD OF ALL CASES OF THEFT FROM CARS, THE IN-CAR ENTERTAINMENT (ICE) EQUIPMENT WAS THE TARGET

It's bright and shiny and easy to sell in the bar tonight. Don't imagine that your £500 stereo will fetch a good price, either. Most units change hands for just enough to make it through to the next hangover. Apart from obvious precautions (coded & removable sets, etc.) there's good old common sense - sticking 2ft square speakers on your rear shelf, and covering your rear screen in stickers advertising the equipment manufacturer, simply saves the thief the trouble of finding which car has the goodies!

THEFT FROM VEHICLES WAS 61% OF ALL RECORDED VEHICLE CRIME (EXCLUDING VANDALISM)

And on an unprotected car, it's so easy; most cars can be opened by a professional thief in a matter of seconds. It takes just a few more seconds to relieve you of your stereo, tapes, handbag, etc.

ALARMS DO WORK AS A DETERRENT

Usually difficult to measure, an official report suggested that more than 80% of convicted car thieves would not bother attacking a car equipped with an alarm. They also said that an alarm siren being activated whilst they were attacking a vehicle would cause them to run away.

42% OF CARS WERE FITTED WITH CENTRAL LOCKING BUT ONLY 19% WERE FITTED WITH DEADLOCKS - THE SINGLE MOST EFFECTIVE DETERRENT

Central locking is often considered a luxury, but its contribution to vehicle security should not be underestimated. The professional thief can spot someone in a rush a mile away, and that person is more than likely to forget to lock a door or two or maybe

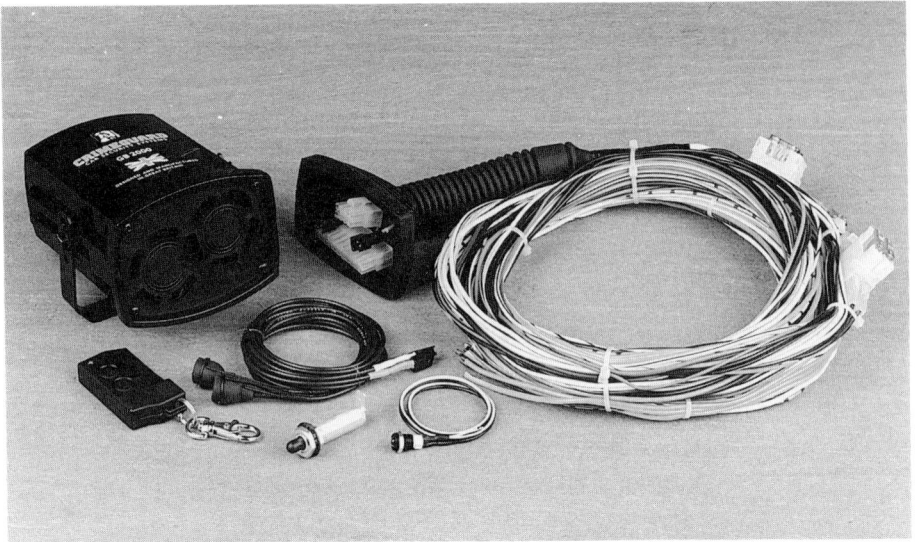

Official figures suggest that 80% of car thieves would avoid a car fitted with an electronic alarm system such as this. The large unit to the left is the siren. (Courtesy Electrosystems Ltd)

Central locking is a very worthwhile security measure. You can buy a kit like this to convert your car and ...

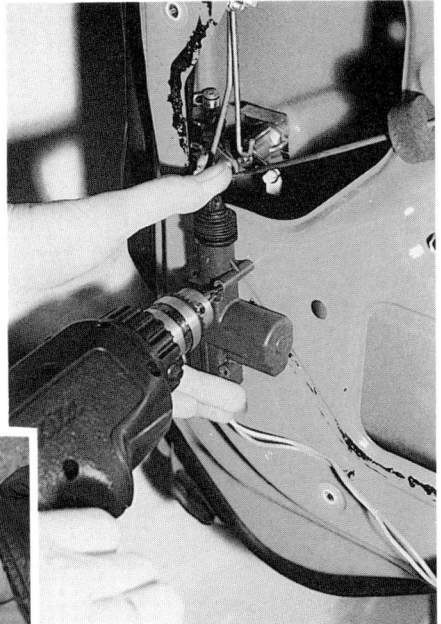

... such kits require no more skill to fit than an average car stereo system, though DIY fitment of some tailgate locks is tricky.

Thieves attack ordinary door locks with great force, which the locks can't withstand. This brings us to ...

... the deadlock. If your car's doors are secured by deadlocks, the average thief won't be able to open them. In a recent comparitive test of car security a skilled thief entered most cars in less than 30 seconds! The only ones which presented a real problem (a GM and a VW) were those fitted with deadlocks. (Courtesy Vauxhall Motors Ltd)

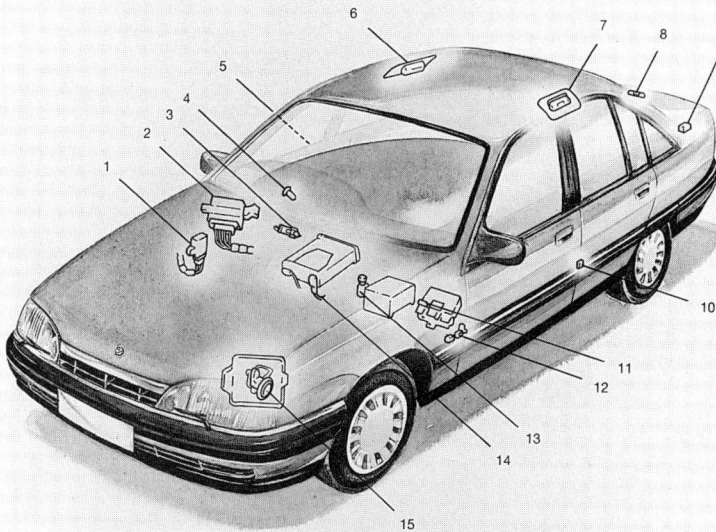

Car manufacturers are moving in the right direction. This photo shows a GM car with standard fit anti-theft devices, including ultrasonic sensors and deadlocks. (Courtesy Vauxhall Motors Ltd)

1 Anti-theft system control unit, 2 Central locking system control unit, 3 Diagnostic plug, 4 Ignition lock, 5 Door lock, 6 Ultrasonic sensor, 7 Ultrasonic sensor with LED, 8 Boot/trunk lid lock, 9 Boot/trunk lid contact, 10 Rear door contact, 11 Relay, 12 Front door contact, 13 Bonnet/hood contact, 14 Radio contact, 15 Siren.

the boot/hatch. And there's nothing easier to open than an unlocked door! With central locking fitted, the chances of forgetting to lock a door are drastically reduced. Aftermarket central locking kits are available for most cars and are easy to fit.

AROUND 65% OF STOLEN CARS ARE USED FOR 'JOY-RIDING'.

A misnomer if ever there was one, for taking a vehicle for a few hours in this manner brings anything but joy, often not even to the thief. Sometimes, the cars are used in other crimes (as getaway vehicles etc.) and often they are involved in accidents with other vehicles, with subsequent injury to thief and innocent motorists/pedestrians. Whilst cars stolen for this reason are usually recovered, they are frequently crashed, burned or totally vandalised.

MORE THAN A QUARTER OF ALL CARS STOLEN ARE STOLEN BY 'PROFESSIONALS' WHO THEN BREAK THE CARS TO SELL FOR SPARES OR CHANGE THE CAR'S IDENTITY BEFORE SELLING IT ON.

So, having paid good money for your car, you're going to give it away (at your own expense in terms of insurance etc.) and will have to buy another one. As well as the usual security devices, a car with its registration number or VIN (Vehicle Identification Number) etched into *all* its glass will be far less valuable to the 'get rich quick' merchants.

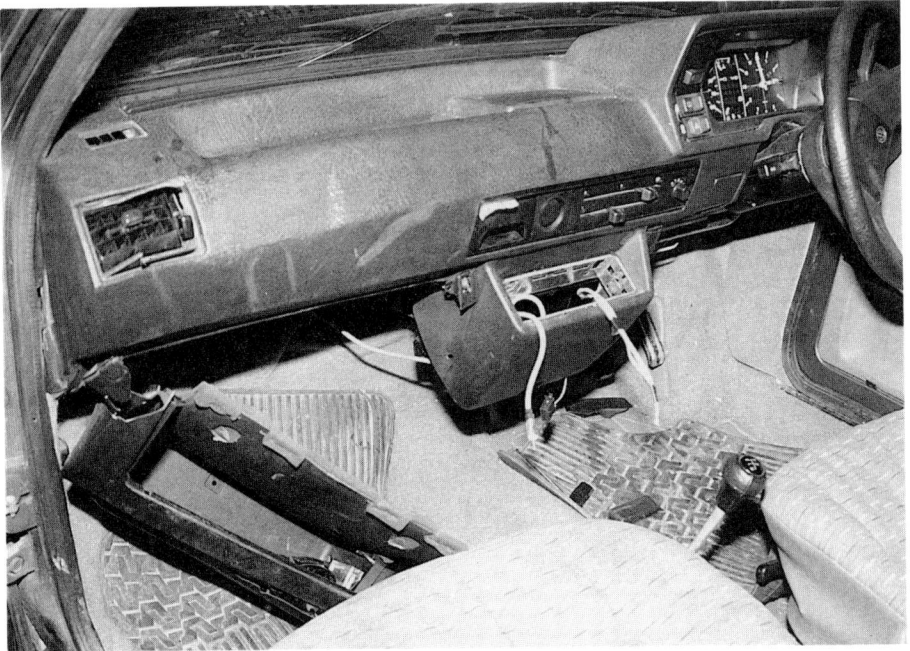

The wrecked interior of the car shown on page 5: audio equipment gone ...

As you can see here, little finesse is used as the thief makes do without the car's own ignition key and switch.

Glass etched with the car's registration or VIN number is a good idea. Some etching service companies also offer a security register system. (Courtesy Retainacar)

❏ THE COSTS

In the UK alone the estimated cost of putting right all this car crime is around £100 million every year. That's a lot of cash by any measure of means. It's tempting just to count this as insurance losses; but of course, what insurers pay out this year, you pay out next year in extra premiums!

❏ SUMMARY

If those statistics don't worry you, you must live inside Fort Knox! Even if you live on a small island or a remote mountainside, there's no room for complacency - for that is the car thief's greatest weapon. The belief that 'it can't happen to me' is the single biggest reason that it happens to so many people - every day of every week of every year. Though it's tempting to think that all this fuss about car security is just a lot of hoo-

WARNING

THIS VEHICLE IS LINKED TO THE NATIONAL VEHICLE SECURITY REGISTER AND 24 HOUR OWNER/MILEAGE CHECKING SERVICE
TEL: 081 871 1333

hah created by the alarm industry to get you to buy their products - it isn't. There is no doubt that faced with a choice of tackling a car with some security device fitted and one without, the thief will take the latter every time.

All this statistical information leaves us with a final fact; in the UK, on average, a car is broken into *EVERY 20 SECONDS*!

HELP YOURSELF
TIPS AND ADVICE ON PROTECTING YOURSELF AND YOUR CAR

Fitting your car with a number of security systems and locking devices can do nothing but good, as long as they are fitted correctly. However, there are many ways in which you can help yourself and prevent the thief from helping himself ... and they're free!

Remember, the thief is essentially an opportunist; a shy, lazy creature, fond of the dark and a lover of the easy life. If you can make your car unattractive to him, he'll move on to easier pickings.

❏ LOCK IT
29% OF DRIVERS ADMIT TO LEAVING THEIR VEHICLES UNLOCKED AT SOME TIME (10% REGULARLY!)

Of course, it's obvious. You lock your car whenever you leave it. But do you? Every single time? When you leave your car for a few minutes only is the time your car is most likely to be stolen or have something stolen from it. Try timing yourself whilst you open the

passenger door, reach in and remove something from the glovebox. You'll find it takes only seconds, so by the time you come back from your 'quick errand', you could find you're missing valuables from your car. As already recommended, central locking makes it easier to ensure that you don't accidentally forget a door.

❏ PARK IT 1
75% OF THEFTS OCCUR DURING THE EVENING OR NIGHT, WITH 70% ACTUALLY TAKING PLACE DURING THE HOURS OF DARKNESS.

The thief doesn't want to be seen, so parking in busy shopping streets or where your car is overlooked by offices and shops is a good idea. Leaving your car in a dark, dingy alley or narrow street is just asking for trouble. At night, keep near the street lights and away from dark, secluded corners. The same principle applies to multi-storey car

You hold the keys to your car's security in your hand. Locking the average car door will slow the thief down and could be the difference between driving or walking home! Always remove the ignition key when leaving the car, even if you are only going to pay for fuel or the car's locked in your garage.

parks, where your personal safety can often be as much at risk as that of your car.

PARK IT 2

55% OF THEFTS ARE FROM CARS PARKED AT OR NEAR THE OWNER'S HOME
32% OF THEFTS OCCURRED AT THE WEEKEND.

Psychologically, you feel safe when you get home: and most people are at home over the weekend. However, it doesn't take a mathematical genius to connect these factors to the statistics above. At home, try to find room in the garage between the kids' toys and the freezer to get the car in as well.

If you can't garage your car and have to park it on the street or on your driveway, take the same security precautions you would in an inner city car park; lock it, alarm it and bring all valuables inside; carphone, radio/cassette deck, cassette box, briefcase, etc. No thief is going to risk breaking into a car that doesn't look as if there's something worth stealing inside. The thief depends upon your 'safe at home' feeling to make his illegal living. If anything, your vigilance on home ground should be greater than ever.

Sometimes, it doesn't pay to advertise and the de-badging of prestigious and/or sporty models is a growing trend in an effort to throw the car thief off the scent.

BADGE IT

In a curious about turn, it is becoming fashionable and infinitely sensible to make your car as anonymous as possible. Many owners are removing the model badges from their vehicles, or replacing them with badges from lesser models in order to fool the thief into passing by.

Some carry this further into the area of exterior specification, where such items as alloy wheels are replaced with steel wheels and body addenda are dispensed with altogether. As these bright, shiny, go-faster bits and pieces appeal to the Magpie nature of the thief, it's certainly a good idea, even if your pride and joy doesn't look quite as 'trick' as it might. The owners of de-badged cars often take quiet delight in the 'Wolf in sheep's clothing' approach.

The same principle can be applied to manufacturer's decals and window stickers. You really shouldn't tell the thief about all the wonderful accessories you've fitted - the advertisment will only whet his appetite and cause him to focus on your pride and joy. Far better to let him think your car is boring, standard and not really worth his time and effort.

❏ HAB-IT

Thieves love creatures of habit. If you leave your car in the same place for the same length of time, whether it be every day while you go to work for 8 hours or once a week for an hour's shopping, you are playing into his hands. Some variance of routine puts some element of risk in the game and will, hopefully, make him choose a safer target.

❏ PERSONAL SECURITY

Attacks on car drivers have increased over the years, with women being the main targets and many are planned whilst the driver is away from the vehicle. When the owner comes back to the car, the attack is made. So, it pays to ensure that the potential attacker cannot tell whether the vehicle is driven by a women or a man. A car with this month's issue of *Cosmopolitan*, three lipsticks and a handbag on the seat is likely to be driven by a woman. Another pointer is the positioning of the driver's seat - because women generally tend to have the seat much further forward. For a female driver it's the work of seconds to ease the seat back a couple of notches when getting out of the car and forward again on her return. For male or female drivers, an alarm system with a 'panic' feature is worth considering, especially if you drive/park regularly in urban areas.

The 'mugging at the traffic lights' syndrome is becoming frighteningly common which brings us to yet another advantage of central locking. There are possible drawbacks, like being trapped inside the car in the event of an accident (most systems are designed to unlock upon impact and if you're not sure about your car check its handbook or with the manufacturer), the odds still stack-up heavily in favour of it being a good idea to lock all the doors once the car is on the move. Central locking makes this easy - especially if you are alone.

The falling cost of cellular 'phone rental has made it much more realistic for them to be fitted purely as an aid to personal safety, and the major motoring organisations are pressing home this very point. The number of tragic incidents, particularly involving women drivers travelling alone or with young children, make such emergency 'phones well worthwhile considering. (Courtesy Celinet)

❏ SUMMARY

✔ ALWAYS LOCK YOUR CAR WHEN IT IS BEING LEFT - THAT'S ALL DOORS AND BOOT/TAILGATE *AND* PETROL CAP.

✔ ALWAYS REMOVE THE IGNITION KEYS AND MAKE SURE THAT THE STEERING IS LOCKED - EVEN IF YOUR CAR IS IN YOUR GARAGE.

✔ PARK SENSIBLY - WHERE THERE IS LOTS OF LIGHT AND LOTS OF PEOPLE.

✔ EITHER REMOVE VALUABLE ITEMS OR PLACE THEM OUT OF SIGHT, PREFERABLY IN THE BOOT.

✔ NEVER LEAVE CREDIT CARDS (AND/OR A CHEQUEBOOK), OR OFFICIAL VEHICLE DOCUMENTATION IN THE CAR.

✔ ALWAYS ARM YOUR ALARM SYSTEM WHEN YOU LEAVE YOUR CAR - HOWEVER QUICKLY YOU MAY INTEND TO RETURN.

✔ IF POSSIBLE ALWAYS REMOVE YOUR STEREO EQUIPMENT. IF NOT, DISGUISE ITS PRESENCE OR FIT A SEPARATE SECURITY DEVICE.

✔ RETRACT YOUR CAR'S AERIAL (ANTENNA) - IT'S AN EASY TARGET FOR VANDALS.

❏ PERSONAL SECURITY

✔ DON'T PARK IN DARK, SECLUDED PLACES.

✔ MAKE SURE SOMEONE KNOWS YOUR DESTINATION AND YOUR ESTIMATED TIME OF ARRIVAL.

✔ IF YOU HAVE A PORTABLE 'PHONE, MAKE SURE THAT THE BATTERY IS ALWAYS CHARGED.

✔ IF YOU'RE A MEMBER OF A MOTORING ORGANISATION, CARRY YOUR CARD WITH YOU.

✔ LOCK YOUR DOORS WHILST DRIVING, ESPECIALLY IN BUILT-UP AREAS - CENTRAL LOCKING MAKES THIS EASIER.

✔ TAKE NOTE OF ANYONE HANGING AROUND A CAR PARK - THEY COULD BE SIZING UP WHICH CAR/PERSON IS WORTH ATTACKING.

✔ FIT AN ALARM SYSTEM WITH A PANIC BUTTON, KNOW HOW TO USE IT AND DON'T BE AFRAID TO!

✔ LEAVE NOTHING IN THE CAR WHICH COULD INDICATE YOU ARE A FEMALE.

THE RIGHT SECURITY SYSTEM FOR <u>YOUR</u> CAR

CHOOSING THE SECURITY DEVICES/SYSTEMS FOR YOUR NEEDS

If you've got this far, then you'll probably be thinking that your car might need some kind of protection. Then again, you might still be coming up with some of the same old arguments. By far the most common of these is -

❏ "MY CAR'S NOT WORTH ALARMING ..."

If your car's value is doubled when you top up the engine oil, then you may be right. But for most of us, our car is the second largest value item we purchase and at any given time represents a large proportion of our personal assets. If you're buying your car on credit (and most people are) then you'll get a reminder every time you look at your bank statement as to just how much your car is costing you.

But the actual value of the car is only the start. What about alternative transport? If you car is stolen some distance from your home, you've got to get back. And if you've a partner and/or kids to consider, it could cost a bundle just to get back to square one. When you get there, you'll then realise how much you depend on your car. You'll have to

This gleaming white Porsche 911 Speedster demands more security measures than most. But don't be complacent, the car thief doesn't look solely for high-powered exotica; from Lada to Lamborghini, every car's a target.

get to work, the shops, the schools ...

The inconvenience factor is substantial though often ignored until it happens. If you have to hire a car, you'll soon discover that it is not a cheap operation. If you have to use public transport, you will discover that it, too, is not cheap and neither is it particularly handy. Then, there's the insurance side of things.

Claim forms are always best completed with professional help, and are the best possible reason for ensuring you're insured with a reputable company - not necessarily the cheapest in terms of premium. Their help at this stage is priceless as it's so easy to give an answer which ruins your chance of successfully claiming off another party.

❏ THE COSTS

You may not think that the actual street value of your car is much, but what about the excess on your insurance policy - the amount of the claim *you* have to pay? There will be a theft and a damage excess. Older drivers, with 'normal' cars and clean driving records, will usually have very low excesses. However, a young driver of a performance car living in a major city could well have an excess equal to a month or more's wages - if he or she can get insurance! And don't forget your no-claims bonus/discount; for most claims, you can lose 2 years' worth - probably a substantial amount ...

If you've ever made an insurance claim, you may have come across the phrase 'uninsured loss'. This is as it sounds - you have a loss which is not covered by insurance.

In many cases, the contents of your car are not insured (or are in-sured only for a small amount) and neither are your legal costs in claiming from a third party. With regard to your car's audio system components, unless you specifically requested uprated cover, you will probably find that you are covered only for the equipment provided by the manufacturer as standard - or not at all. Radios, tape players and CD units are a favourite target for the thief, hence the insurers' reluctance.

As most of us buy our cars on credit, it is common to find that the amount offered by the insurers isn't enough to pay off the loan, particularly where there is a penalty for early redemption.

❏ CONVERTIBLES

Almost by definition, owners of convertible vehicles find themselves highly rated: most convertibles are valuable, or sports models, or a combination of both. Though convertibles are not that much of an insurance risk (compared to a tin-top variant of the same car), they are always easier to get into - there aren't many steel-roofed cars that can be opened by using a penknife! A good quality alarm system and extra-special vigilance when parking are the best pieces of advice on offer.

❏ THE CONTENTS

Even if a thief agrees with you that your car isn't worth stealing, he could still do a lot of expensive damage (windows, door locks, paintwork, dashboard) breaking in to get at the valuables you left for him. Stop and think: where is your car parked now and what is in it? What could a thief steal if he broke into your car whilst you were reading this? A case full of

RISK ASSESSMENT CHART

How to use this chart. Determine which risk groups your car and area fall into and then cross-reference both in the "Risk Category" chart. The resulting code letter represents the level of security you should consider.

THE CAR

"TOP RISK"

⇨ Sports and high-performance cars. Convertibles. Vehicles of high value.

"HIGH RISK"

⇨ High performance medium-sized cars, top-range executive saloons, modified cars.

"MEDIUM RISK"

⇨ Popular cars up to 4 years old.

"LOW RISK"

⇨ Cars over 4-years old and/or those with little intrinsic value.

These definitions are generalisations and should be used as such. There are obvious anomalies (a 10 year-old Ferrari would warrant something more than a steering wheel lock!) and of course, no-one knows better than you if your car is likely to attract a thief by dint of its modifications or fitted accessories. A 6 year-old Vauxhall Nova would attract more attention than normal if it were fitted with £1500 worth of BBS alloy wheels and a pair of Recaro seats at £600 a throw! Adjust your grouping accordingly - tend toward a higher group if you're unsure, in order to be on the safe side.

THE AREA
(where your car spends most of its `risk' time)

"TOP RISK"

⇨ Inner city areas. Large urban housing estates, particularly where crime is prevalent and/or where there is a high incidence of unemployment.

"MEDIUM RISK"

⇨ Town centres, car parks (especially multi-storey), overnight street parking - particularly outside your own home.

"LOW RISK"

⇨ Rural areas, any quiet village locations.

Note for commuters: If you live in the country but work in the city, take the latter as `your' area. If you're not sure about the area in question, contact your local police station or crime prevention office, which will be able to tell you how various areas are risk-rated.

tapes or CDs? A good quality radio/cassette deck with speakers and amplifier? A handbag with this week's wages in it? Credit cards? A boxful of tools?

Don't think because the contents of your car aren't worth *that* much, a thief won't bother. Many thieves are happy to take the value of to-night's beer money. If it causes £500 worth of damage to your car, then that's your problem. Having your car stolen or broken into is a very painful, expensive and inconvenient experience. It's up to you to do as much as you can to prevent your own suffering.

❏ THE ANSWER
So where do you start? How do you know which security systems or devices you should be fitting to your car? The chart on page 18 will help you decide, though it should only be used as a guide (there's always the exception ...) and you should not hesitate to fit extra security measures if you think fit.

❏ SUGGESTED SECURITY OPTIONS
See "Risk Category" chart-
A. The 'works'. A top-quality electronic alarm system with battery back-up, protecting all points of entry with some form of interior sensing. A pager is an option worth considering. Some form of immobilisation (ei-ther built-in or as an add-on). Passive arming highly advisable.
B. A high-quality electronic alarm system, but without some of the 'whistles and bells' such as pagers. Immobilisation still important.
C. A mid-range electronic alarm system, possibly used in conjunction with some form of physical security.
D. Physical security a priority (*eg:* steering wheel/gearlever locks). Should be brightly coloured and easily visible to the potential thief. Contents theft is more likely than car theft, so a portable interior sensing device is recommended.

Note: Certain security measures should always be taken, regardless of the vehicle or area. For example, alloy wheels should be protected by a set of locking wheel nuts/bolts and a removable radio/cassette deck should always be removed.

❏ COUNT YOUR LOSSES
How much would it cost you to have your car stolen? The following list details the basic categories of loss and expense. Some of these costs may be covered by your insurance policy (in which case you can feel smug about your wise choice) but, unfortunately, most drivers will find themselves adding together fairly large amounts as they go through the list. Compare the total losses with the cost of a good security system - in

RISK CATEGORY CHART

AREA ▶ VEHICLE ▼	TOP RISK	MEDIUM RISK	LOW RISK
TOP RISK	A	A	A
HIGH RISK	B	B	C
MEDIUM RISK	B	C	D
LOW RISK	C	D	D

many cases, particularly where a high-performance car and/or inner city rating are involved, the theft excess alone will warrant the expenditure.

✗ HOMEWARD BOUND
If your car is stolen whilst you're away from home, you'll have to get back. If it's any distance, you'll have a fair old taxi bill, especially if you've wife/husband/friend/kids, etc. in tow.

✗ BUSINESS TRAVEL
Can you get to work (and back!)? If you can't would you lose any pay? What about the loss of commission in a sales job?

✗ HOLIDAY TIME
An up-and-coming holiday could be seriously jeopardised by the theft of your car. Taxi and/or rail fares for a family would add massively to the cost.

✗ COMMUNICATIONS
Phone calls, letters and travel involved in sorting out your claim soon mount up.

✗ CAR HIRE
Do you need to hire a car? Check out the local rates; even for a tiny runabout, it's far from cheap. If the police find your car severely damaged, but repairable, you could find yourself waiting well over a month for its return.

✗ HOME ALONE
If you leave your house keys in your car, you've given the thief *carte blanche* to raid your house. Replacing a set of door locks isn't cheap.

✗ INSURANCE - THEFT EXCESS
Do you have a theft excess? Most do. It's the first part of any claim that you have to pay. It can be considerable, especially where the car is highly rated and/or the driver is young and/or it is a high-risk area. Check out your personal combination and then your excess.

✗ INSURANCE - NO-CLAIMS DISCOUNT (NCD)
If you claim, do you lose your no-claims discount? If so, it's usually 2 years' worth - that's probably 20% on the next premium and 10% on the one after that.

✗ INSURANCE - CONTENTS COVER
Are the contents of your car covered? If so, for how much? A common victim here is car hi-fi, sometimes not covered at all, it is often the subject of a small maximum limit.

✗ INSURANCE - UNINSURED LOSS
Insurance policies contain lots of words, many of which describe what *isn't* covered. As well as your contents, there could be plenty of other costs not covered, including car hire, medical costs, legal costs (often considerable) and accommodation away from home. It's well to ask *before* you take out a policy exactly what is and isn't covered. A cheap policy is not always the best one in the long run.

CHOOSING ELECTRONIC SECURITY SYSTEMS

WHAT TO LOOK FOR & THE VALUE OF INDIVIDUAL FEATURES

Follow the advice given in Chapter 3 and particularly the chart on page 18. However, take into account your personal circumstances. For example, you may live in a quiet, rural area and drive a small, basic car so, technically, it doesn't warrant a high security rating. However, whilst the car's book value may be small, its value to you may be enormous; not only in cash terms (loss of insurance discount, excesses etc.) but also in sheer inconvenience if the car was not available. And in some cases, a car could literally be a lifeline. Look at the 'count your losses' section in the previous chapter to see how you fare.

Choosing the best electronic security products for your individual needs can be difficult and most of us have to work within a budget, which limits our choices. Buying a basic alarm which is suitable for expansion in stages is a way to spread costs without compromising quality. Recommendations from existing users are always helpful, given that they come from a reliable and unbiased source. Many motoring magazines (and consumer magazines) run tests comparing like with like and give unbiased opinions on individual systems.

❏ SIREN

Pick the nosiest one you can. The choice is complicated, as the method of rating (in decibels - dB) is not standardised and there is legislation in the offing to create a maxi-

Sirens come in various shapes and sizes. Left is a standard siren complete with the electronic 'works', in the centre is a smaller one which could be used as an extra siren (or as the main siren in a modular system where the electronic brains are located inside the vehicle), and at right is a siren, with electronics and battery back-up.

mum limit. At present, a 'real' dB figure of between 115 - 130 db is seriously loud.

Early 'compact' alarms were prone to unreliability as moisture and heat inherent in the engine bay affected their delicate electronics. However, progress has ensured that, in terms of reliability, there is now little to choose between modular and compact systems.

Battery back-up is highly recommended. On a system not so fitted, the siren can be silenced by simply cutting the battery lead. Professional thieves know how and where to lay

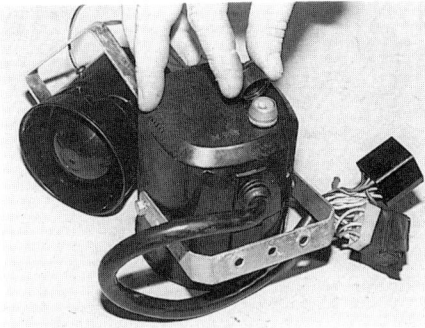

Systems with battery back-up should feature a key-operated off switch in case they develop a fault.

under the vehicle in order to snip the cable and deactivate the alarm before it can even begin to do its job. Doing this on a system with its own battery back-up will actually cause the alarm to trigger and the siren to sound, after which it will reset and continue to protect the car. The siren/control unit houses ni-cad batteries which (usually) take a trickle charge from the vehicle to ensure that they are always ready for action. There must be a method of legitimately switching off the alarm, usually by means of a key in the siren unit.

❏ POINT OF ENTRY PROTECTION
The alarm should be triggered by the opening of the doors, bonnet (hood) and boot/hatch (trunklid/tailgate). Depending on the system you choose, you may need to buy extra pin-switches. For a vehicle with no interior protection (ultrasonic sensors etc.) this is essential; for a vehicle *with* interior protection, it may be belt and

All points of entry should be protected. Here a pin switch is being fitted.

braces but is still recommended.

❏ VOLTAGE DROP SENSING
This type of sensing used to have a very bad reputation, being one of the greatest causes of false alarms. Where this function is used, the alarm will trigger when it detects electricity being used in any of the car's electrical systems. The theory is that the alarm will trigger should, say, the courtesy light come on as a door is opened. The problem is that many cars have functions which legitimately use power after the ignition key has been removed (*eg.* radiator fan, cellular 'phone). The other main voltage drop bugbear was that of temperature - as temperature drops, say, during a cold night, the voltage in the vehicle battery also falls. On some systems, the alarm would sense this as an intruder. Again, more false alarms. Many alarm systems are now offering 'intelligent' voltage drop

sensing, whereby the electronics are able to distinguish between a telephone call and a burglar. They usually also realise that the gradual depreciation in battery voltage caused by a decrease in temperature is nothing to get excited about. Less intelligent systems usually offer the option to switch off the voltage drop sensing and discretion is often the better part of a good night's sleep!

❏ INTERIOR PROTECTION

Whilst pin switches will cause the alarm to trigger if a door is opened, they don't help if a window is broken and the thief reaches into the car to steal the contents. That's why you need some form of interior sensing device. Most popular are ultrasonic sensors,

Sensors come like this as two separate sensor heads which are placed at the top or bottom of the A-pillar (the vertical pillar at each side of the windscreen/ windshield). These need to be wired to a control interface mounted beneath the dash. Alternatively, multiple sensors can come in a one-piece pod which can be mounted on or under the dashboard or, on occasion, on the shelf behind the rear seat.

Philips remote control alarm with battery back-up and Tri-sonic sensors (which measure the size of an intruding object, its speed and the ambient temperature). Multi-sensing should prevent false alarms triggered by flying insects or changing temperatures. (Courtesy Philips Car Systems)

though they can be a little sensitive to wind pressure.

For an open-top vehicle, sensors are required which are not affected by air pressure (so the hood can be left down) and which can 'see' through plastic trim (so they can be mounted where a thief cannot reach them). At present, this means either Microwave or infra-red sensors or 'hybrid' sensors, such as the Philips Tri-sonics or the Digital Microsonics. The last two claim to combine the best of both worlds, with microprocessor-controlled, intelligent sensing.

You will often find that interior sensing can be added as an optional extra after the main alarm has been fitted, enabling you to spread the cost and ease your cashflow. Which options are available is something to check *before* you buy your system!

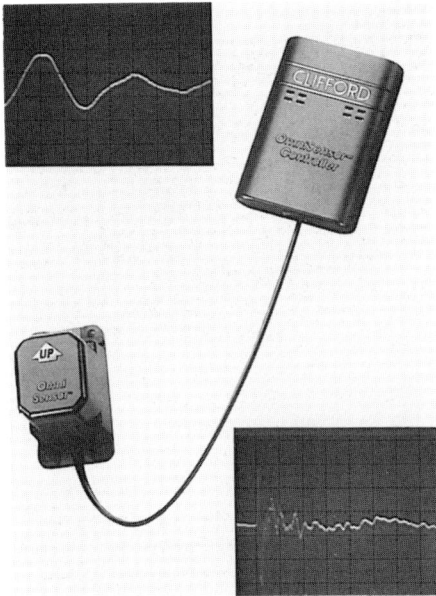

The major American company, Clifford, have a range of sensory devices reckoned to be at the forefront of security technology. This photo shows their Omnisensor, which they claim to be the world's only all-electronic, microprocessor-controlled user programmable sensor: ordinary shock sensor, this is not! Having set the required sensitivity level, the sensor reacts to more than just the minimum level; it analyses the amplitude, frequency, time constraints and other aspects of the 'shock' signal being detected.

Many alarm systems offer the facility to arm the system but switch off the interior sensors, a good feature if you are leaving animals in the vehicle.

❏ PASSIVE ARMING
Many top range alarm systems feature passive arming, whereby, if you don't actively set the alarm using the remote key, the system will arm itself. (This function is also available on many immobilisers.) It is ideal for those of a forgetful or lazy nature. The alarm monitors all openings (doors/hatch, etc.) and at a set period after the last one has closed (typically around 45 - 60 seconds) it arms the alarm. However, if you go back to the car anytime before that period has expired and open a door, the 'countdown' would stop. Naturally, there is a link to the ignition switch for practicality and as a safety precaution, most passive systems will not operate the central locking/windows etc., (where applicable).

On returning to the car, disarming is carried out in the normal way with the remote 'key'. An alarm which sets itself takes some getting used to, and most owners forget at least once (usually to much embarrassment in a public car park!). Nevertheless, it is

The tilt sensor can be part of the shock sensor or, as here, a separate unit. Its object is to trigger the alarm when it senses that the angle of the car is changing. This is a defence against thieves lifting the front of the car to tow it away or wheel thieves jacking the car.

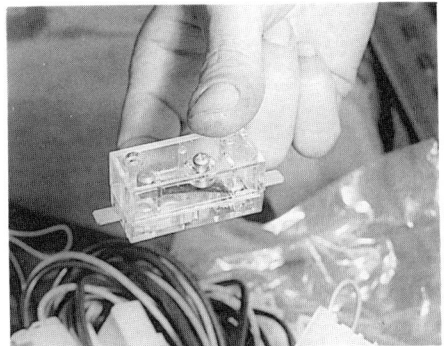

worth giving it a whirl for a week or so (most systems offer the option to delete the passive function); the one time you forget to set your alarm could be the one time the thief picks on you!

❏ SHOCK SENSOR AND TILT SENSOR

The shock sensor is often (though not necessarily) included as part of a siren unit and mounted under the bonnet. It's job is to detect sudden shocks to the vehicle, such as a brick being used to break a window or damage being inflicted on the body-work. Correct sensitivity adjustment is essential to ensure that it does not raise a false alarm every time next door's car passes by.

❏ STATUS MONITOR

This is a feature usually to be found on the higher price range alarms. Most use the warning LED to flash a series of coded messages to the owner on his or her return to the vehicle. This code denotes the last of the alarm functions to have been triggered during the owner's absence. This is particularly useful where a false alarm problem is occurring, as otherwise it is not always possible to tell which alarm function is in error.

❏ MANUAL KEY, DELAY OR REMOTE CONTROL?

Where the choice is yours, choose remote control every time. Key oper-ated systems are relatively easy to defeat and known for their unreliability. Similarly, a system with an entry delay offers the thief an opportunity to deactivate the ap-paratus. There are two types of re-mote alarm; infra-red and radio. Infra-red usually operates only 2/3 metres

Radio remote control systems must be legal and approved alarms usually carry a sticker saying as much. The range is typically around 10 metres, but can be up to 30 metres. An aerial/antenna is employed, which is often simply a piece of wire leading from the system control unit (usually in the siren) though, sometimes, it's fixed to the windscreen/windshield.

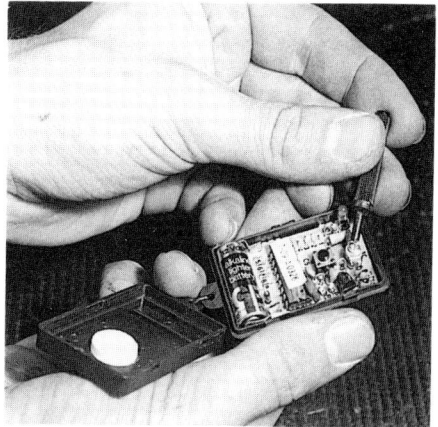

Though they may look delicate, the innermost workings of the average remote handset are actually quite robust. Many can be 'trimmed', as here, in order to obtain the best performance/ range.

from the vehicle and the beam can be deflected by bright sunlight on the vehicle glass.

Some systems offer remote operation as an optional extra, to be plugged-in like interior sensing. Worries over machines used by sophisticated criminals which can 'scan' or 'grab' remote control instruction codes have been largely stamped on by the introduction of sophisticated circuitry within most modern alarms which make them virtually impossible to defeat.

PANIC MODE

The 'panic' option is little-publicised

way to make getting the bonnet open incredibly difficult.

PAGERS

Pagers are a good idea, for as the number of electronic alarm systems in use increases, it is getting ever easier for the public to ignore a shrieking siren on the grounds that it's probably a false alarm. With a pager system, you will be bleeped or flashed by means of a portable paging de-

Pagers need to be of officially approved standards. Check before you buy. (Courtesy Fanfare Electronics)

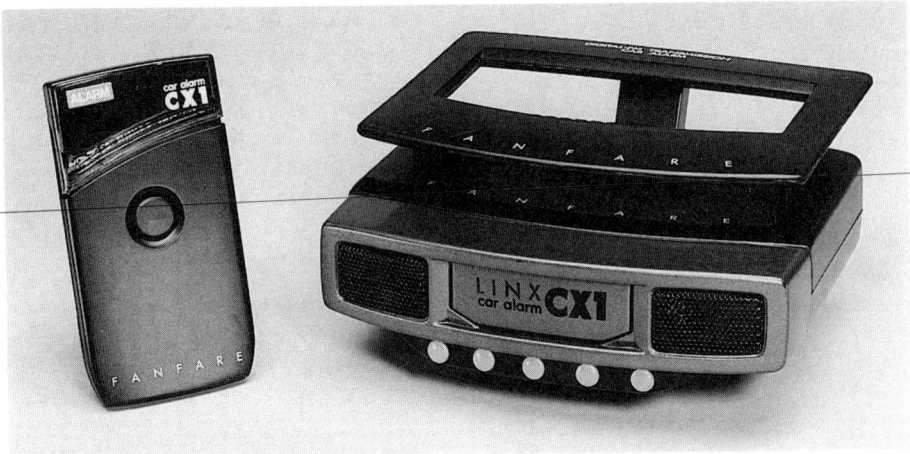

but increasingly useful when attacks on motorists themselves are increasing. When fitted, the remote control handset can be used to instantly trigger the siren (and indicators where applicable), regardless of whether or not the alarm is set.

BONNET/HOOD LOCK

If a thief is to attempt to defeat your alarm system, the simplest method is to attack the siren. Using a bonnet lock wired into the alarm system is a

vice when your alarm has been triggered. The range depends on what is between the car and the pager. In town, expect around 200 metres (220 yards) but in the open the signal can travel up to three times that distance. Pagers are not cheap and are usually only found on top-flight alarm systems for expensive cars.

TELEPHONE LINK

As in-car cellular telephones become more common, it's not surprising to

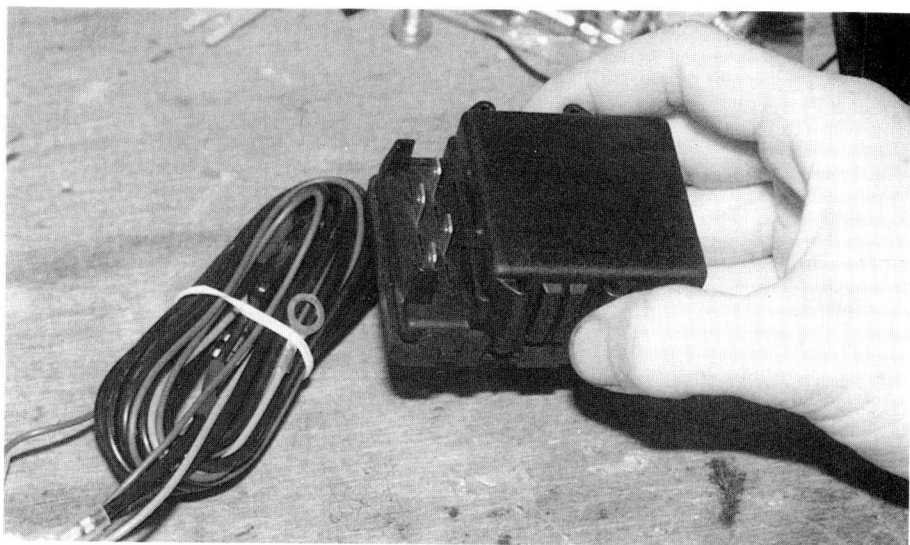

A typical interface - a black plastic box with some terminals on it. Wiring should present few problems but you must be scrupulously neat and methodical. This is especially true if you are fitting your car with a total closure system (electronic windows/doors/ sunroof), as you'll end up routing a myriad of wires.

find that some (usually top level) alarms can be wired to the 'phone. When the alarm sounds, the 'phone will dial a specific number.

❏ OPTIONS

One very important factor to bear in mind is that of optional extras: accessories and/or add-ons which can enhance your basic system.

Some alarms can be purchased as fairly basic delayed-entry systems, with the option to 'plug-in' an interface converting it to remote-control operation at a later date. Such an option is well worth having. (Courtesy Zemco)

Even though cash may be a problem now, it is often useful to be able to add items such as automatic central locking, extra sirens, ultrasonic sensors, etc., at a later date. Some systems, however, do *not* facilitate such additions and you need to know

that before you buy. Ask the seller if the system you are considering is expandable in the future.

It is increasingly common for alarm systems to be able to operate the central locking, electric windows and electric sunroof where fitted, offering total closure and security at the press of a single button. It's not only good for peace of mind, it's impressive in the car park, too! If your car has these (or some of these) electric goodies, then linking them to the alarm system is highly recommended. When securing your car is easy, you're more likely to do it every time you leave the car.

In order to effect this, you will need an interface for each feature - "interface" being a fancy word for a box of electronic bits and pieces which allows the alarm to talk to other electronic components within the car. Some alarm systems have a built-in interface for central locking, whereby it takes just a couple of connections to complete the required circuit. On other systems, you will need to purchase a separate interface for the functions you require and wire it in as

per the manufacturer's instructions.

The most common combination is to use an interface for central locking, one interface for each pair of electric windows and one for an electric sunroof. As you might imagine, even with one interface, there is quite a lot of wiring to be carried out and routed neatly behind the trim panels. It also means that you will need somewhere to mount the interface itself. Always keep separate interfaces inside the car, preferably under the dash area, out of harm's way.

❏ CENTRAL LOCKING

If your car hasn't got central locking as standard, it might be a good time to think about adding it. There are proprietary kits available to operate 2 or 4 doors. On some cars, it is also possible to include the boot/trunk or hatch/tailgate, but it can often be tricky to fit the motor so that it works in total harmony with the rest of the system.

❏ SECURING CLASSIC CARS

Whilst defining a classic car is a thankless task, for the purposes of security it really means cars that are something over 15 years old or more. Modern classic cars (a Porsche 944 or Ferrari 348, perhaps) are likely to pose few problems, as the manufacturers themselves have a range of security products designed to

A proprietory central locking system kit.

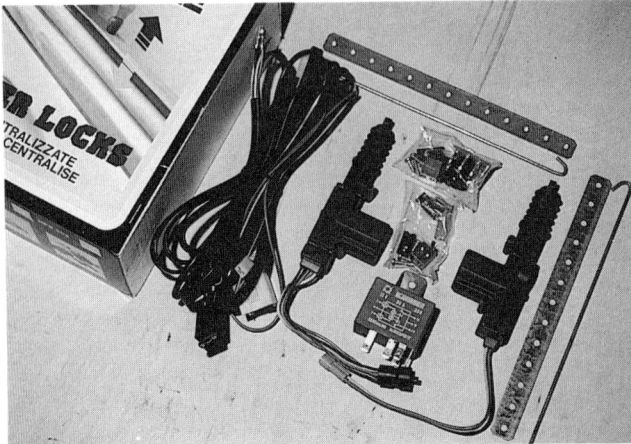

suit. Fortunately, the classic car price bubble of the late 1980s has burst in a big way, meaning that their overall desirability for the thief has declined accordingly; but *not* altogether!

As statistics show, complacency is the car thief's main ally so you should be vigilant at all times. Some form of security presence is better than nothing and the 'Help Yourself' rules detailed earlier in this book apply just as much to a 1967 Rover as a model 25 years younger.

Once we get back to the '60s the problem of positive-earthing (grounding) of electrical systems is likely to rear its ugly head. If you have such a car and want to secure it electronically, then you're well advised to have a chat with your local security specialist and/or one of the major manufacturers.

Where the car has a negative earth (ground), it is ironical that the (relative) simplicity of early engine bays makes many models more suitable than their modern equivalents for fitting an electronic alarm. With plenty of space to use, there's no problem in finding a home for the siren, etc., usually one of the major stumbling blocks with the average modern car. You don't have to look far to find that the common-or-garden courtesy light switch is a fairly modern phenomenon; again, you may be faced with making more holes to utilise certain electronic alarm features.

Naturally, once you start drilling holes and modifying the car, originality suffers as a result. The balance here is often between having a slightly unoriginal car in your garage and a very original car on the police stolen vehicle list.

If your car is *really* something special and a regular concours entry, then you don't need me to tell you that the inclusion of a siren and a clump of extra wiring looms in the engine bay could mean the difference between a trophy and a wasted journey. If your car is *that* good, and you don't keep it locked away for eleven months of the year (you're in a minority, here), you will either have to bite the bullet and lose a few concours points or be extremely careful where you leave your pride and joy - suffering that awful disease, parking paranoia, into the bargain!

Physically securing a classic car is not usually difficult, with a good many 'universal' products being suitable. Portable noise makers/vibration sensors may also be a good bet, although those operating from cigar lighters may be a no-go area in all but the top-of-the-range classics.

If you have a classic, the odds are there's a club for the model and, where there's a club, there are usually lots of well-informed, friendly people, all eager to help you with your security problem. It is well worth the joining fee for the advice and contacts you will acquire.

❏ STICK-ON DEVICES

There are a number of windscreen stick-on devices (illustrated overleaf) like the Auto Alarm . In this case, it is essentially a shock sensor with a siren included, which is stuck to the windscreen/windshield in a prominent position. The flashing LED alerts the thief to its presence and, with the siren being around 130 dB, there's plenty of noise when it goes off.

Lots of racket in the car is going to be unpleasant for any thief, but it is not too difficult for the professional to

*A stick-on elec-
tronic alarm unit.*

this nature is dependent on professional fitting of the equipment.

❏ WHERE TO BUY

The choice here is largely between a specialist alarm (and/or audio) dealership, car

overcome some of these stick-on alarms. These devices usually offer good value for money, but cannot offer the same level of protection as more sophisticated (and expensive!) systems. We would recommend that such devices are not used as the only protection on high-risk vehicles.

❏ WARRANTY

Very important. If the alarm is made well enough, it will have a good warranty. Check it out thoroughly, both in terms of length (currently varying from 1 year to life!) and of how the warranty is implemented. Some warranties require defective parts to be returned to the manufacturer for replacement. In some cases, this could mean waiting for weeks - during which time, of course, your car is not alarmed. Look out for special warranties, such as the one offered by one manufacturer in conjunction with a major motoring organization; if your car (fitted with the manufacturer's alarm, of course) is immobilised due to an alarm fault or due to damage caused by an attempted break-in, the motoring organization could be summoned to assist, free of charge. Not surprisingly, any guarantee of

parts and accessories shops or by mail order. If you are justly confident that you know exactly what you want, then you may (but not necessarily) save money by buying from a large car parts and accessories store. However, you're unlikely to get much in the way of technical and product advice there. On the other hand, if you're not so sure about what you need, the specialist dealer should be able to fill in the gaps easily and recommend specific systems to suit your pocket, your needs and your car. At the end of the day, the one thing you must avoid is a snap decision or impulse buy.

❏ FITTING

Some products require professional fitting in order to validate the warranty. Others are sufficiently complex to make DIY fitting a non-practical proposition. Do not attempt to fit any alarm system unless you are justifiably confident in your abilities. Vehicle immobilisation as part of an electronic alarm system should be wired professionally, or checked over by a professional as soon as you've installed it. Purpose-made electronic and hydraulic immobilisers should always be fitted by a specialist.

❏ BUYING CHECKLIST

✓ Choice. There is a huge choice when it comes to buying security, so don't rush in. Check out performance, warranties and price before you buy.

✓ Which devices you choose will depend on many factors, but remember that your best line of attack is usually to fit more than one security product.

✓ If a system really is good, the seller should be able to put you in touch with a satisfied customer before you buy.

✓ Legality. As new legislation is introduced, check before you buy that the equipment in question is not obsolete and outside the law. For instance, in the UK remote alarms should be DTI (Department of Trade and Industry) approved with a sticker on the unit to prove it.

✓ Independent magazine tests are an excellent method of finding out how good a particular security device really is.

✓ Warranty. What is covered, for how long and how is cover implemented? There are some great guarantees about, so there's no reason to accept a poor one.

✓ Insurance discounts. Some devices carry a premium reduction with certain companies - as long as the item is professionally fitted. Check what (if anything) you gain before you buy.

CHOOSING PHYSICAL SECURITY DEVICES
ALL YOU NEED TO KNOW ABOUT SECURITY HARDWARE

There are other ways to secure your car apart from or as well as full electronic systems. You may choose to use one or more of them instead of or as well as a conventional electronic alarm system - it's hard to have too much security. Remember that whilst a physical device may stop a thief taking your car (or at least slow him down) it won't protect the car's contents from theft. The first aim of any physical security device should be to attract attention - you want the potential thief to notice your security measures and walk away. The opportunist thief (the most common variety) does not want a hard time which increases the risk of being caught.

Most products take the form of hefty lumps of metal, usually hacksaw and bolt-crop resistant. They are designed to slow down, if not stop altogether, the casual crook. At the time of writing, there are devices which will -

☆ Prevent the steering wheel from turning more than a few degrees.
☆ Prevent the clutch and/or brake pedal from operating.
☆ Prevent movement of the gearlever.
☆ Prevent release of the handbrake (parking brake).
☆ Prevent the road wheels from turning by physically locking one or more wheels with a clamp.

The Stoplock 2 is a hefty device which fastens around the rim of the steering wheel and makes it impossible to turn the wheel to drive the car - even if the steering lock is overcome. The device's own lock is unlikely to be defeated as it is a radial pin tumbler type, with more than 10,000 possible combinations. The device totally obscures the steering wheel central boss so that a thief cannot remove the steering wheel to get around the Stoplock. Made of high grade steel with a soft, grey outer covering to prevent damage to the vehicle interior.
(Courtesy Metro Products Ltd)

The Eagle Claw is so named for the way it grips the sides of the steering wheel. It is made from toughened steel, with a unique ratchet design so that it fits most vehicles. Ring claim that the high-security keys supplied are virtually impossible to duplicate and they also offer a cash-back guarantee should your car be stolen whilst fitted with the Claw. Like the Stoplock, it covers the centre of the wheel and is large and bright enough to be a deterrent to a potential thief. (Courtesy Ring Automotive)

If the Krooklok seems familiar, that's because it is - it's been around for 27 years! It comprises a steel shaft which locks around the steering wheel rim and the brake pedal. Fluorescent hooks ensure a high visibility factor. (Courtesy Krooklok Products)

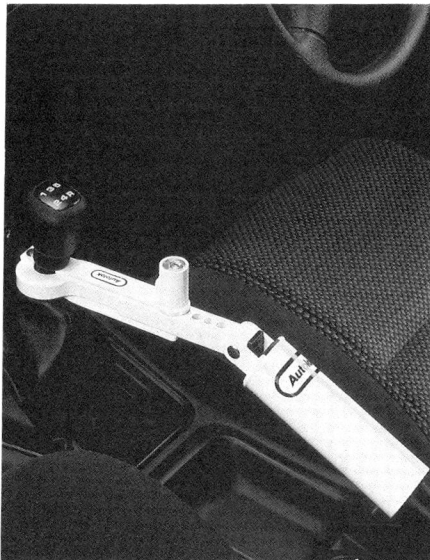

The Autolok device seeks to prevent vehicle theft by making it impossible to release the hand-brake or use the gearlever. It slides over the handbrake and is telescopically adjustable so that it will reach any manual or auto-matic gearlever. Its bright yellow finish ensures high visibility and the makers claim that the 5000-combi-nation lock is tamper proof. (Courtesy Autolok Security Ltd)

33

Left. The Sonix alarm locks around the steering wheel. It's a bright red plastic box containing a siren and ultrasonic sensors. As such, it protects the interior of the car, taking its power and charging its internal nicad batteries from the cigarette lighter socket. If the plug is removed from the socket whilst the device is armed, the 125dB siren will sound, and inside the car that's some noise. Its portable nature means that it can be used from car to car.
(Courtesy AAA Security)

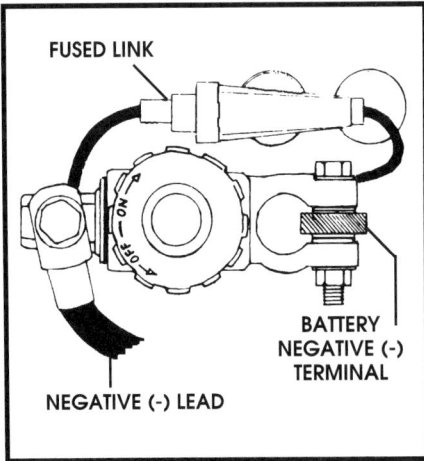

FUSED LINK

BATTERY
NEGATIVE (-)
TERMINAL

NEGATIVE (-) LEAD

Above. The Dis-car-nect is a simple device. It replaces one of the normal battery terminal con- nectors and, once the knurled plastic nut is removed (and popped into your pocket) the car's main electrical power is cut (a small fused link continues to supply power for alarms, clocks, coded radio/cassette units, etc). The device is not hugely difficult to overcome, but it does prevent a thief just diving into your car and 'hot-wiring' the engine. It is also quite handy to have during gen- eral maintenance of the car.
(Courtesy Richbrook International)

Below. The sight of a wheel clamp may cause many of us to cringe, but when it's self-fitted, it is an effective way of protecting your car. There are several on the market, all of which work in the same way. Though fitting a clamp in wet and cold weather could be inconvenient, it is certainly very visible and difficult to overcome. This particular model has a selec- tion of holes through which can be inserted a padlock for a 'belt and braces' approach. And for those with caravans/trailers, this kind of protection is ideal. (Courtesy Klamp-it Innovations Limited)

Right. Alloy wheels are always a good seller when it comes to stolen goods. A set of locking wheel nuts or bolts costs a fraction of the cost of a new set of alloys, takes hardly any time to fit and require no maintenance. CAUTION! You must ensure, though, that the bolts (where applicable) are exactly right for your car and the cone-seating dimensions of wheel bolts or nuts are also a crucial safety factor. You must also remember to keep the locking key in the car, so that you can remove a wheel in the event of a puncture. And there's no room for complacency if you've got steel wheels, for the price of tyres makes just about any wheel/tyre combination worth stealing.

Below. Some thieves won't bother with your car or its contents, they'll just relieve you of your expensive fog or driving lamps. Using locking nuts, similar to those used to pro-tect road wheels, will help you to keep your lights. Some electronic alarm systems have an accessory function, whereby such items as fog/spot lamps can be included in the protection.

Above. Many aftermarket sunroofs can be popped open by a deter-mined thief anxious to get at your car's contents. Opting for a roof with a locking mechanism, like this one from Automaxi, is a good idea.

☆ Remove the battery power eas-ily.

☆ Lock your car's alloy wheels in place.

☆ Lock your foglamps in place.

With only a few exceptions, such products are designed to be univer-sal in application and require no actual 'fitting'. This means that they can be transferred from car to car. Shown in this chapter is a selection of physical security products, just a few of the myriad now available.

The information in this chapter will help you to make a good choice of device type but quality varies, so be guided by consumer tests or personal recommendations if possible.

❏ CARAVANS/TRAILERS

The cost of modern caravans/trailers makes them well worth stealing. They are much easier to resell than cars and very difficult to trace. Your first line of defence is to store your caravan/trailer somewhere safe. If you are using a commercial caravan/trailer park, check what security procedures are followed and what kind of insurance is operational. Securing your trailed unit once it's away from the relative calm of the secure park starts by ensuring that it is hard to move. This means either locking the towing mechanism or locking the whole lot to the towing vehicle.

A wheel clamp, as described and shown earlier, is a good idea, as is removing the unit's wheels if it is to be parked for some time. Some car alarm systems (Digital, for example) have an 'accessory' feature which allows the alarm to be triggered by the unplugging of the trailer plug/socket. Such complex alarms aren't cheap, but then neither are caravans!

It is possible to have a conventional electronic alarm fitted to your caravan, though you will need a reliable power source. Using standalone items, such as the Sonix unit, featured earlier, is another alternative.

The Metro universal anti-theft coupling lock is a simple but effective device which fits all 50mm trailer sockets. The unique combination key releases hardened steel pins which hold the lock in the socket. It comes with a storage bag which makes it easy (and clean!) to store when not in use. (Courtesy Metro Products Ltd)

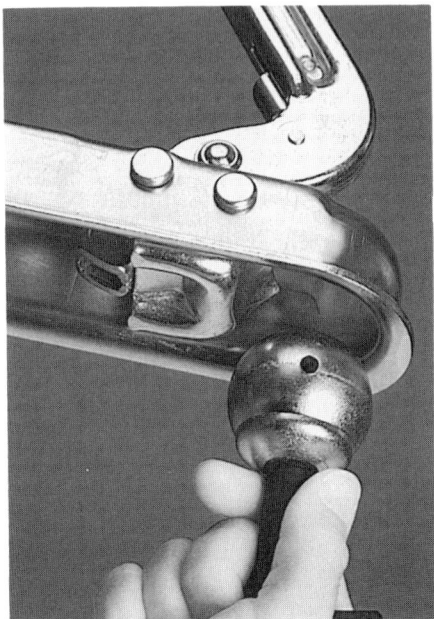

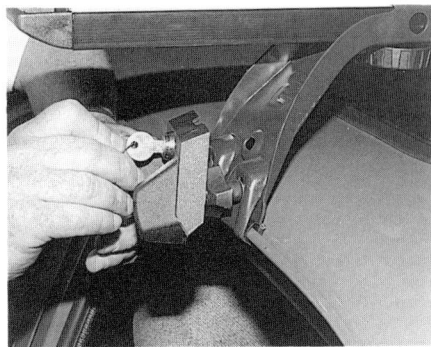

Many bars can have locking covers added as an optional extra: opt for this extra.

❏ ROOF-RACKS AND ROOF-BOXES

It is becoming increasingly common to use sophisticated roof-racks/boxes to carry bulky excess luggage. Like the car itself, these will need to be secured, not least because the racks themselves are often expensive items.

Aerodynamic and cavernous roof-boxes such as this one are incredibly useful, but can expensive. Make sure that there is some way of locking your possessions in the box, the box to the rack and the rack to the vehicle - you can't be too careful.

TOOLS AND TECHNIQUES
THE TOOLS YOU'LL NEED AND WIRING TECHNIQUES

Fitting an electronic alarm system yourself is not necessarily a difficult process, but it is one which must be completed in a thoroughly professional manner. Some alarms are simpler to fit than others, a fact which is usually made clear in the labelling and, to a certain extent, in the price. It is important that you know your limitations and capabilities and stay well within them. If you are fitting an alarm in order to gain an insurance discount, check first; it is probable that professional fitting (with receipts to prove it) is a pre-requisite.

It is essential that you have enough time to complete the job (allowing for a spot of fault-finding and alarm adjustment), the right tools and the right accessories. A wrong connection in an alarm system could cause you inconvenience (as it false-alarms at 2am) or worse, danger, as your engine immobilises at high speed.

❏ TOOLS
Listed here are the tools required to fit a standard alarm system in an average car (* optional) -

Crimper/wire stripper
*Soldering iron ***
Electrician's tape
Dot punch
Socket set/spanners (wrenches)
Flexible grab/coat hanger
Petroleum jelly (Vaseline)
Test light/multimeter
Terminals/insulators

*Cored solder ***
Electric drill & bits
Masking tape
Screwdrivers
Rust preventative
Copper-based grease
Tape measure/rule

Alarm connections should be soldered wherever possible. If you can't solder, then it's a good time to learn - but don't use your alarm system for practice. Crimping comes second to soldering as a way of making good electrical connections and you're better off with a good crimped join than a bad soldered one!

Make sure you have a good supply of terminals to suit whichever method you choose. Some have insulators built-in (usually coloured as to the amperage capacity, which makes it harder to make a mistake) and some use separate insulators. Whichever, don't leave terminals uninsulated. Use electrician's tape rather than nothing. You should also be aware that the amount of preparatory work involved in fitting even a modest alarm system is considerable, particularly the removal and modification of interior trim. All the trim has to be replaced on completion: another reason for allowing plenty of time.

❏ WIRE
The stuff used to link all those clever electronic bits and pieces together.

The kind of wire you'll be dealing with in alarm fitment comprises a number of thin copper strands held together and insulated by a plastic sheath. Wire terminology revolves around how many strands are contained in the sheath and how thick each is. For instance, "14/0.012" denotes a wire of 14 strands, all 0.012inch thick (the measurement might be metric).

The more current passing through the wire, the thicker it has to be. Alarm manufacturers equip their products with wire to suit, so if you have to extend a wire for any reason, make sure you use the same thickness - and colour.

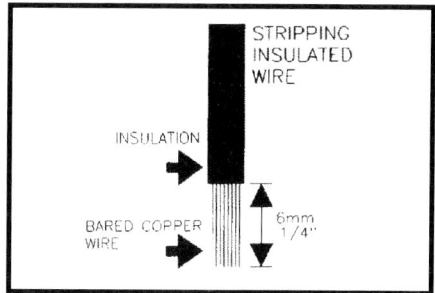

When making a wiring connection, strip around 6.5mm (1/4in) of the plastic sheath from the end of the wire and twist the copper strands together. Make sure you don't cut any of the copper strands, otherwise you'll reduce the current carrying capacity of the wire.

SIZE OF WIRE (IMPERIAL)	CURRENT RATING (AMPS)
9/0.012	5.75
14/0.010	6.00
36/0.0076	8.75
14/0.012	8.75
28/0.012	17.50
SIZE OF WIRE (METRIC)	
9/0.30	5.50
14/0.25	6.00
4/0.30	8.50
21/0.30	12.75

off, you also risk electrical overheating, as the terminal is overloaded. It's a good idea to buy a 'selection box' from your local auto-spares shop to start with, replacing individual types/ sizes as and when necessary.

Lucar (or spade) connectors - *Probably the most common for in-car use. Seen here, right to left, are a male, a female and a piggy-back. Using the latter is essential where you need to take more than one wire to a single point - make sure, however, that you don't overload it.*

❏ CONNECTORS AND TERMINALS
Use the right type of connector for the job in hand: there are a large number of sizes and types. The larger the current, the larger the wire diameter and the larger the connector must be. Pushing a very thick wire into a connector which is obviously too small is asking for trouble; not only is there every possibility it will come

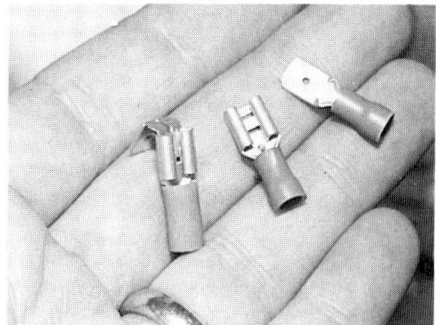

Bullet connector -

Aptly named and useful for joining leads where disconnection may be required in the future. They can be soldered and/or crimped.

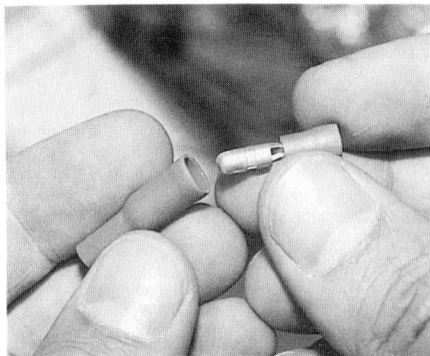

Butt connector -

Joining two wires in a more permanent manner can be achieved by using a butt connector.

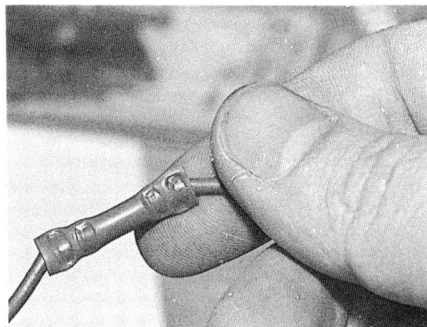

Scotchlok -

The ubiquitous Scotchlok connector is a God-send for the non-electrician, however, they do have their limitations. Certainly, it cannot be recommended that they be used in the engine bay (where moisture and heat could cause problems) and they should never be used in alarm power wires and earth ground wires or in immobilisation circuits. Use them for the indicators, if you must. Prefer-

ably, take a little time and learn how to make good crimped or, better still, soldered joins.

The principle of the Scotcklok connector is fine, as it uses a sharp metal cutter to slice through the plastic sheathing of the two wires to be joined and makes an electrical contact.

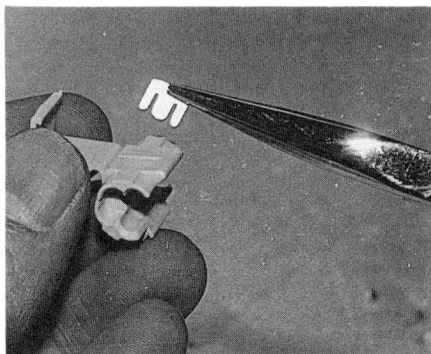

The 'trade' tends not to use Scotchloks for essential circuits because of the unreliability which can occur if the plastic sheath is not fully cut through or where the cutter is too vicious and actually damages the copper strands. Note that these connectors come in various colour-coded sizes - use the right size ...

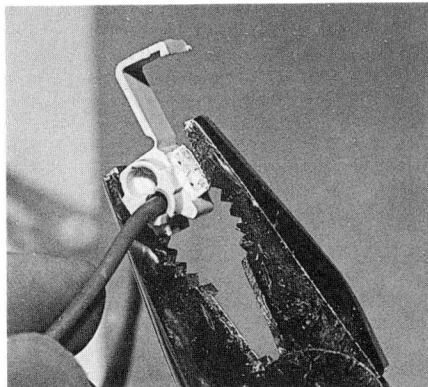

❏ ELECTRICIAN'S TAPE

This is an excellent way of creating a loom from a number of wires being routed in the same direction and of insulating connections. DO NOT, however, twist two wires together, cover them with electrician's tape and call it a join; it isn't, it's a bodge. At best, it will be inadequate, at worst, it could be downright dangerous. DON'T do it!

Another way of routing a number of wires around your car's engine bay is to use professional (multi-wire) looming cable. This is available from auto-electrical specialists by the metre. Not only will it give a tidier appearance, it makes life harder for a thief, when he's looking for which wires to cut.

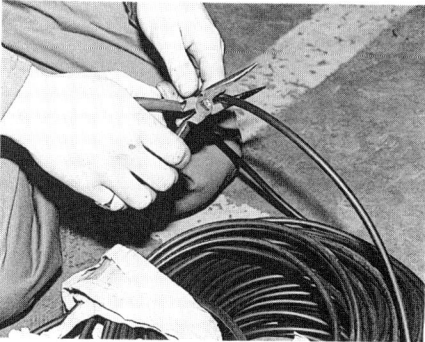

❏ FUSES

All cars are fitted with a fusebox and a selection of fuses. A fuse is the weakest link in any electrical chain and its purpose is to protect a more valuable piece of electrical equipment and, ultimately the car. Should a malfunction occur, the fuse is designed to sacrifice itself for the common good, but in order to work correctly, it must be a suitable rating for the job in hand. Ratings of both applications and fuses (see table) are

Shown immediately below are the most common fuse types in use today. Left to right: ceramic, flat-bladed and glass. As shown in the photo (bottom right) alarm looms almost invariably contain one or more in-line fuses (which could be glass or flat-bladed). Most modern vehicle fuseboxes use flat-bladed fuses and older cars utilise the ceramic type. Before you start fitting your alarm, make sure you have a few relevant spares handy in case a problem causes one of them to blow.

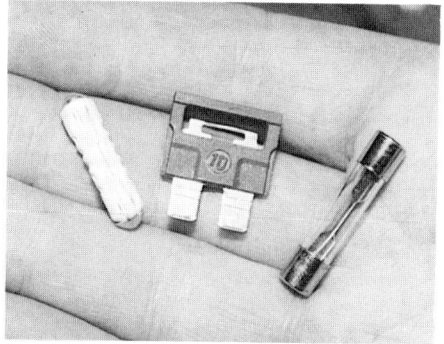

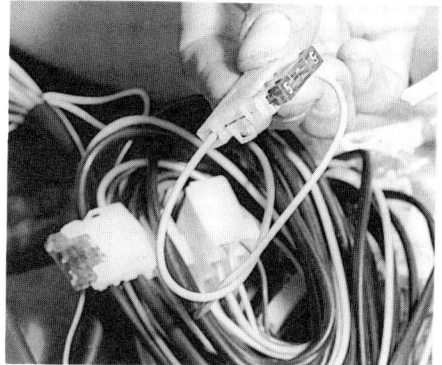

denoted in amps. NEVER change a fuse rating - either up or down. NEVER substitute anything else - bits of old wire, cigarette paper, etc - for a fuse, not even on a temporary basis.

FUSES

FLAT-BLADED FUSES	COLOUR CODE
3 amps	PURPLE
4 amps	PINK
5 amps	ORANGE
7.5 amps	BROWN
10 amps	RED
15 amps	BLUE
20 amps	YELLOW
25 amps	WHITE
30 amps	GREEN

CERAMIC FUSES

5 amps	YELLOW
8 amps	WHITE
16 amps	RED
25 amps	BLUE

GLASS CARTRIDGE FUSES (TUBULAR)

3 amps	BLUE
4.5 amps	YELLOW
8 amps	BROWN
10 amps	RED/GREEN
35 amps	WHITE

GLASS CARTRIDGE FUSES (FLAT)

2 amps	RED/BLUE
5 amps	RED
8 amps	BLUE/GREEN
10 amps	BLACK/BLUE
15 amps	BROWN
20 amps	BLUE/YELLOW
25 amps	PINK
35 amps	WHITE
50 amps	YELLOW

❏ SOLDERING

The best way to make an electrical join is by soldering and, ideally, all alarm-related joins should be made this way. However, if you can't solder at all, a good crimped join is better than a bad soldered one!

There are 3 types of iron in common use. A mains-powered iron (top) will (usually) require an awkward extension lead to get it to the car. Also, the 'bits' tend to be a little too big for in-car use. Powered from the car battery, a 12v iron (below) is generally more handy to use and geared more to the requirements of auto electronics.

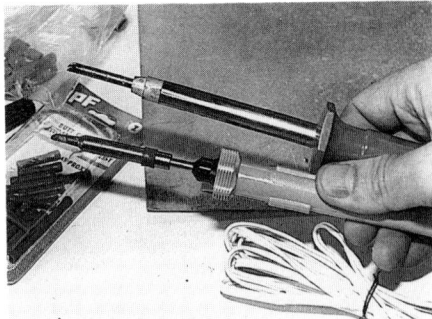

As an alternative, butane gas-powered irons can be purchased as individual items or in a set with a number of different bits, widening the scope for using the tool. They're more expensive, but, being totally portable, are ideal for in-car work.

✠ SAFETY ✠
Caution! All soldering irons run at temperatures above 700 degrees C (1292 degrees F). Naturally, great care is needed, so wear protective gloves, boots and goggles: remember too that most interior trim is highly flammable and so is that tankful of fuel!

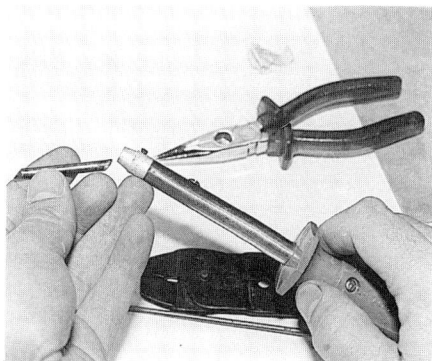

1 Top left. The bit is the interchangeable piece of metal at the end of the iron where the heat congregates. The larger the electrical join, the larger the bit required.

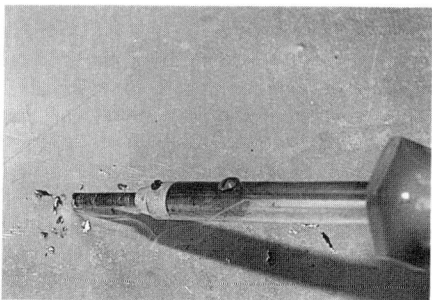

2 Centre left. Before you start, you need somewhere to rest the very hot iron and to deposit any excess solder. Electric irons can be 'tapped' onto a sheet of metal, but a gas iron should be wiped gently on a damp sponge.

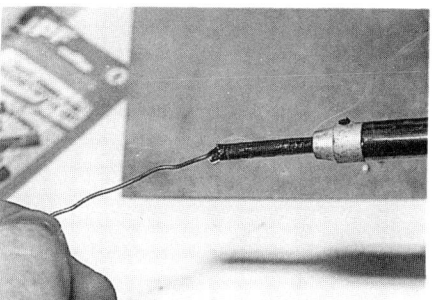

3 Bottom left. If your iron has a new bit, then you should 'tin' it before you start work (ie: melt solder in a thin film over the bit).

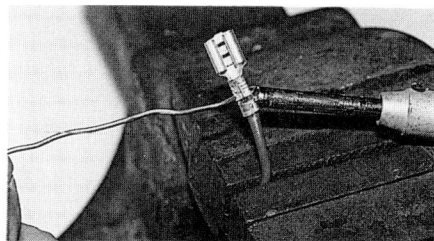

Top right. Always apply the iron to the work-piece BEFORE applying the solder. Allow the work-piece to heat sufficiently for the solder to flow quickly through the join. Once this has occurred, remove the heat. Do not blow on the join (it will cool quickly enough on its own) or allow the join to move before it has set; 'dry' joins could result and cause endless problems with a bad connection.

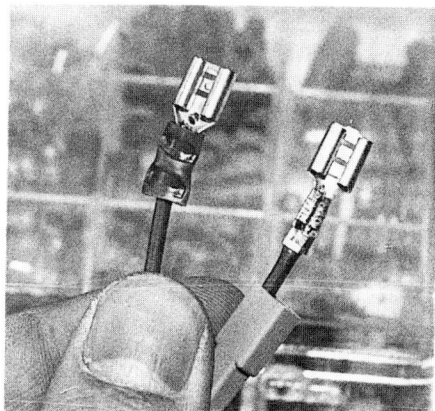

Above. The finished join at right, with a crimped join alongside. Note the protective insulator waiting to be slid over the soldered join.

It's a good idea to make a number of practice joins before you start work in earnest. You'll soon discover that the heat from the iron moves quickly from the join along the wire to your fingers. Solder out of the car wherever possible, with the work-piece held gently in a vice (vise). Failing that, use a pair of pliers or other clamp to hold things in place and to act as a heat sink so that the wire's insultation doesn't melt.

❏ CRIMPING

If you can't solder (or if soldering isn't

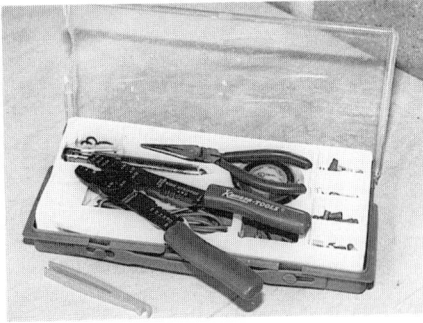

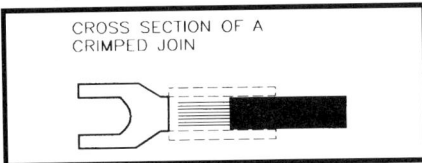

A good way to start your electrical tool kit is buy a set such as this, which includes long-nose pliers, crimpers, test light, fuse-puller and a variety of connectors.

practical) then crimping comes a close second. You'll need solderless crimp terminals, readily available from car parts and accessories shops everywhere.

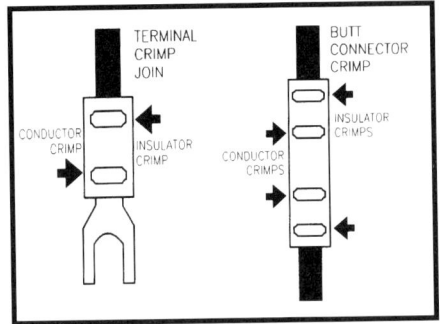

CROSS SECTION OF A
CRIMPED JOIN

Place the bared end of the wire fully into the crimp terminal ...

... (above) and then squeeze hard on the crimping tool. Most crimpers have colour-coded jaws so that the join is crimped hard enough but not too much. Always crimp in two-places; crimp over the copper wire and then again, over the insulating sheath. This 'belt and braces' method means that the physical strength of the join lies in the second crimp and therefore the electrical integrity is not compromised. (Above right). When joining two wires together using a butt connection, you will need two sets of crimps (ie: four in total). When you have finished any crimped join, give the wire a tug to ensure that the join is sound.

❏ DRILLING HOLES

An electric drill will usually be required as you'll need to drill at least a couple of holes. Whether you're drilling in metal or plastic trim, make sure that you won't damage anything behind or beneath the drilling area in question: it's easy to plough through an existing wiring loom or hole a water hose.

When drilling a hole in painted metal, there will always be the danger of the drill bit slipping and damaging the car or, of course, yourself. To avoid this, always place masking tape in a 'cross' over the hole centre, then use a dot-punch followed by a small pilot drill.

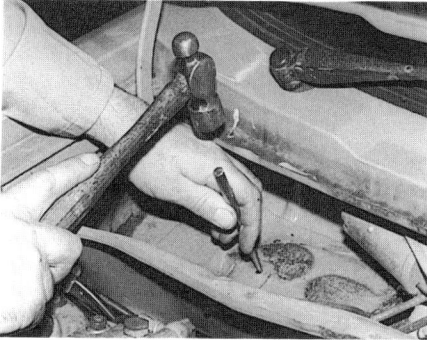

Dot-punch the hole centre before drilling.

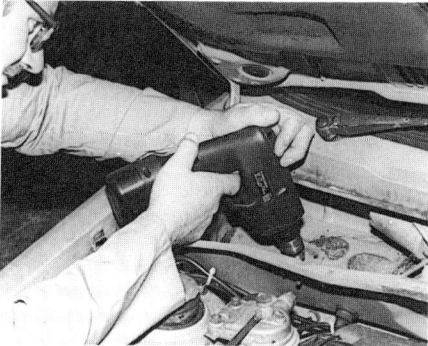

A cordless drill is the best bet for use when working on a car. Not only is there no awkward mains lead but, being smaller, these drills can get into more confined spaces. In addition, the power delivery is softer than most mains drills.

❏ TEST LAMPS AND MULTIMETERS

When you're dealing with your car's electrical system, you will need some way of knowing which wires are live. The simplest way is to use a test lamp, a screwdriver-like device, with a wire connected to the top and a croco-

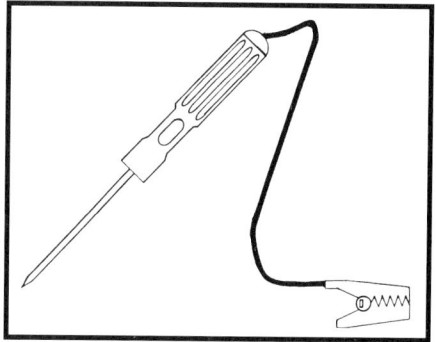

A test lamp-type circuit tester

dile clip at the other end of the wire. The bulb enclosed in the centre of the tester lights-up when power is detected. Most test lamp probes are extremely sharp (and demand respect when working) so that they can be pushed through the plastic sheathing of a wire. This means that wires can be tested part way along their length, rather than at a bared end or terminal. It can be used to test for live wires (connect the clip to a known earth) or earth wires (connect the clip to a known 'live').

A multimeter will achieve the same

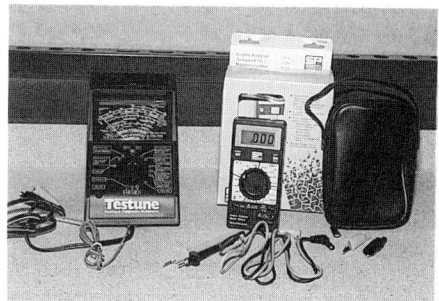

Apart from the make of meter you prefer, the choice is also about style of readout: digital display (right) or a swinging needle (left). Ideally, try both before you part with your cash.

ends in the same way as a test lamp, but offers much more beside. Apart from being able to test resistance (ohms) and various levels of voltage, it offers an exact measure of the voltage. This is particularly useful on many modern (and electrically com-plex) cars, such as some Jaguars, where many wires can be live, but only with 5 volts. Those who regularly work on their own cars should be able to justify the cost of a multimeter and a quality unit will serve for many years, thus spreading the cost.

FITTING AN ELECTRONIC SECURITY SYSTEM

STEP-BY-STEP GUIDE FOR THE DO-IT-YOURSELFER

At present, there are few rules with regard to who can and who can't fit an alarm system. However, there are moves afoot within the alarm and insurance industries to introduce legislation regarding the DIY fitting of alarms and/or immobilisers. The 'ifs, buts and maybes' and general ramifications of this will probably rumble on for some time, but you should make it your business to know the law before you start work. Ask the dealer from whom you buy the alarm; if he doesn't know, you're at the wrong dealer! Anyway, we'll assume it is legal to fit your alarm on a DIY basis and that's the route you've decided to take.

Before you even buy the alarm, give some thought to your vehicle warranty. On new, or nearly new cars, there is often a comprehensive and valuable body/mechanical warranty in place. It is probable that the fitting of an alarm by the car's owner would invalidate it; certainly, drilling holes in bodywork covered by a anti-perforation guarantee in order to fit an alarm system would present you with a difficult case should rusting occur in the vicinity. You should weigh-up the amount saved by self-fitting against the amount you stand to lose in the event of a lost claim. However, it must be said that if it is evident an electronic alarm system has been fitted correctly and with care, then the car manufacturer would have no good cause to claim that the vehicle's warranty is affected.

Whilst some alarms are relatively easy to fit, the top-of-the-range systems require no small amount of skill, particularly when it comes to vehicle immobilisation systems. If an honest appraisal of your technical abilities has you thinking twice, discretion, in the form of a dealer fitment, could be the best option.

❏ BEFORE YOU START

Task one is to open the box and match the contents to the diagrams - the time to find out something is missing is before you start! Study the instructions until you know what goes where and how. This is best achieved indoors, in peace and quiet, coffee to hand. Trying to work everything out in a cold garage, with your car in pieces is not the best way to start.

If, having read the manufacturer's product-related instructions and all of the text in this and the previous Chapter (6), you have doubts about your ability to fit the electronic security system then leave the fitting to an expert, or undertake the task with the aid of someone who is confident.

When all is clear and you're happy to proceed, assemble the tools described in the previous chapter.

Before starting work, the battery earth/ground terminal should be disconnected. IMPORTANT! If your car has a security coded radio/cassette unit, make sure you know the number

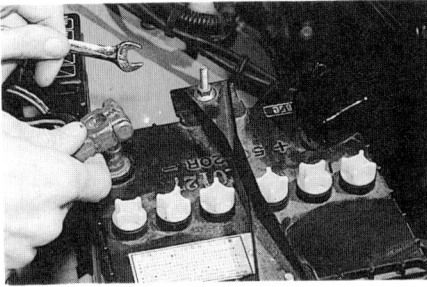

Before starting any other work, disconnect the battery's earth/ground terminal.

before you disconnect the battery - otherwise it could cost you hard cash to get the unit operating again! Also, if your car has a complex computer/engine management system, check with the manufacturers before disconnecting the power; some are very sensitive and some record details of the engine's behaviour between servicing. For certain procedures, you will have to ascertain which leads are 'live'. Ideally, check these first or, alternatively, temporarily reconnect the earth/ground terminal to check.

❏ **THE SIREN**
The siren is usually positioned under the bonnet/hood (or front luggage compartment, if a rear-engined vehicle). If a slave siren is fitted, then this could be fitted elsewhere in the engine compartment or in the vehicle cabin. The latter can be very useful, as it is very difficult for a thief to concentrate on stealing your car or dismantling your ICE system with a 120 dB siren screaming in his ear!

As sirens like to be dry and cool the inner wing/fender or the bulkhead/firewall is a common site for the unit. Some cars have room for a siren close to the wiper motor atop the front bulkhead. This is fine, as long as water cannot drain directly onto the siren.

If your alarm has battery back-up, then there will be an override or 'valet' lock - you will have to allow room to insert the key. In a compact system, the wiring loom is plugged into the back of the siren. Allow room for this and think about the routing of the wiring, as there can often be 20 or more separate wires to distribute. Balance this with the need to make the unit as inaccessible as possible to the determined thief, who may force open the bonnet and attempt to silence the siren and/or disarm the alarm system. It's a quandary which requires no small amount of fore-

Don't position the siren where it will be in a direct spray of water or where water is likely to drip on it. Don't position the siren close to sources of extreme heat.

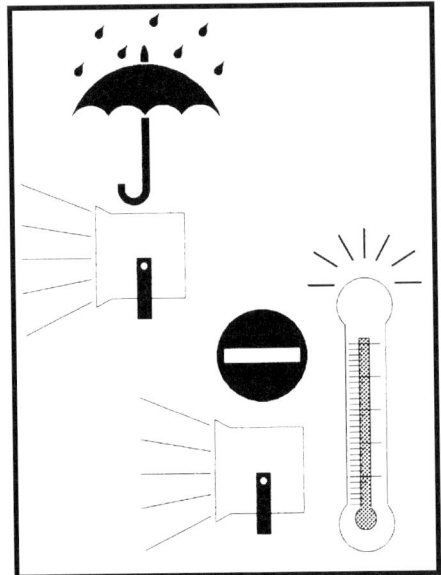

thought, and which usually leads to a compromise solution. It is important that no wiring should pass near sources of extreme heat (such as the exhaust manifold) or moving engine parts (such as the cooling fan and drivebelts.).

When you've decided where to position the siren use its mounting bracket to identify and mark the positions for the mounting bolt/screw holes. Once the pilot holes are drilled, use a varicut (or christmas-tree) drill bit to enlarge the holes to the size required. Remove any rough burrs with a round file and treat with rust preventative to combat the onset of rust around the unpainted metal.

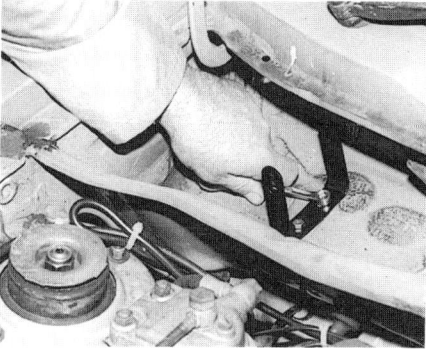

Fit the mounting bracket using the screws or bolts provided. It's a good idea to treat the screw-threads to a dab of grease to prevent them rusting solid. Many fitters would now plug the alarm loom into the siren and fit the whole lot into the car. This is not the professional way, however, which is to deal with the loom as a separate entity before fitting it to the car.

❏ THE ALARM LOOM - DIVIDE AND CONQUER
A little forethought can make wiring

your alarm system so much easier. Straight out of the box, a complex alarm could include a bunch of wires as thick as your wrist. It's a good idea to study the loom with the wiring diagram and divide it up before you start installation.

Lay the loom on the ground or on a bench and get all the wiring straight. Separate those wires that are going to stay in the engine bay, those that will pass through the bulkhead into the vehicle cabin and those which will not be used at all. The latter should be taped out of the way, ensuring that they cannot 'short circuit' to each other, to other wiring or against the vehicle's body.

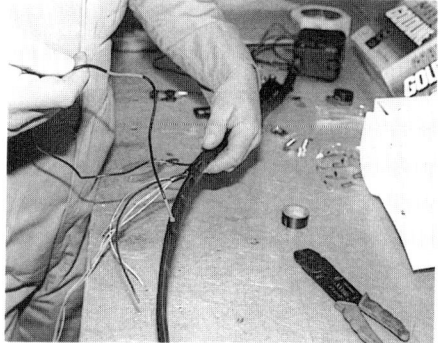

Having sorted out which wires are going where, you can use your electrician's tape to make one or more 'sub' looms.

Wherever possible, it is easiest to strip the insulation from the ends of the wires and fit terminal connectors whilst the loom is out of the car and totally accessible. Think carefully, though, as the fitting of some terminal connectors could make it impossible to thread the loom through holes in the vehicle bodywork ...

Mock-fit the siren on the bracket to ensure that you have the required access and that there is room for the wiring. Then remove the siren and connect the wiring loom multi-plug, smearing the back of the wiring connector (*ie: under* the protective plastic sheath) with a coat of Vaseline which will act as an extra defence against the ingress of moisture.

On this Cobra loom, the multi-pin plug is secured in place by a couple of screws.

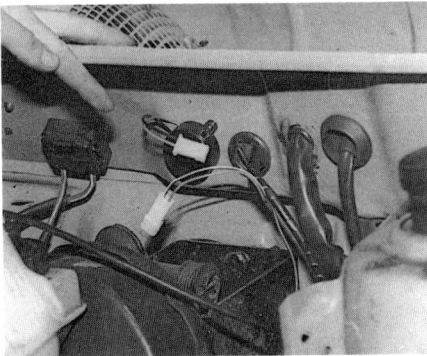

Most alarm systems offer a 'chirp' to confirm that instructions from the remote-control handset have been recieved. On some, the siren is used, on others, as here, a separate 'chirper' is required.

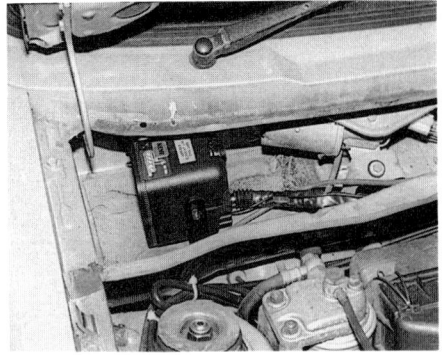

The siren in position, looking unobtrusive and with very neat wiring.

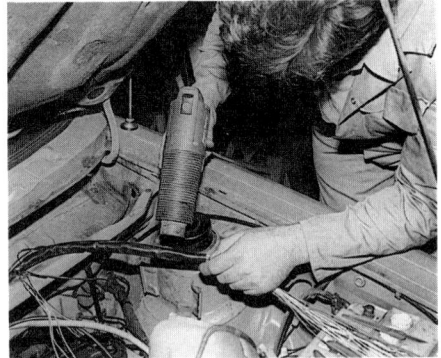

If you can get hold of real loom tape, the application of a little heat will give you factory-look 'shrink-wrapped' wiring. Use a heat gun or hairdryer to shrink the tape. Apart from looking neater, correct looming of wires means that it is more difficult for a thief to identify the alarm wires. Contrast this with amateur fitments, where the engine bay is positively alive with a multi-coloured spaghetti of wiring running every which-way.

When passing the wiring loom through the bulkhead/firewall into the cabin, it is often possible to utilise an existing hole. Whether you do this

Where possible, utilize existing openings in the bulkhead/firewall to route wires into the passenger compartment.

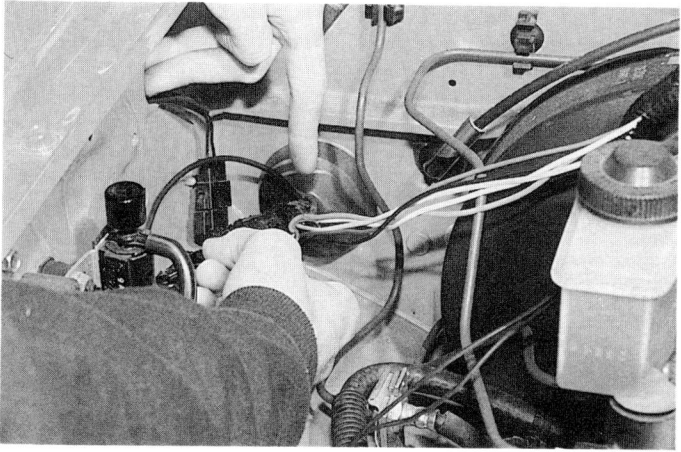

or drill a new hole, it is imperative that a grommet is used to protect any wires passing through a hole in metal. Failure to do so could result in a chafing through of the plastic sheath with subsequent short circuits. If you do have to drill a new hole, make sure that both sides of the bulkhead/firewall are clear and don't forget to paint bare metal to prevent rust forming.

❏ VEHICLE IMMOBILISATION

Many alarm systems come with the facility to immobilise the car via one or more electronic circuits. In essence, it is simply a question of cutting a wire leading to whichever circuit is chosen and placing the alarm system in

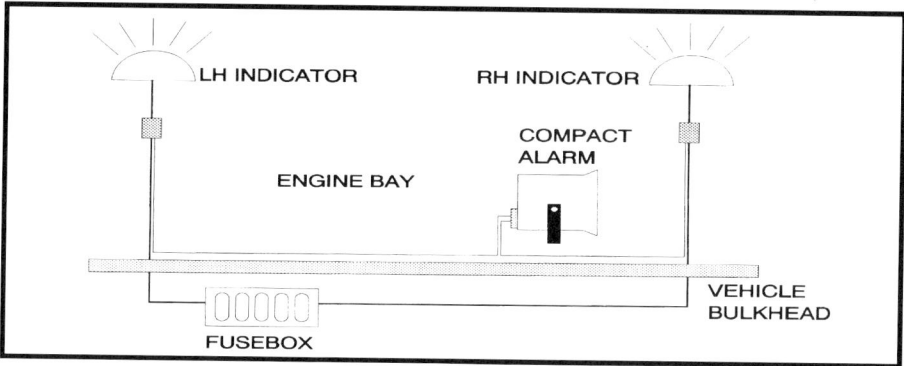

Where the indicators are linked to the alarm, they will usually flash a specific sequence as the alarm is armed/disarmed and operate as hazard lights if the alarm is triggered. There are two alarm indicator wires (usually yellow) and these should be linked into the power carrying indicator wires on the vehicle. This can be carried out under the bonnet/hood, though this can lead to problems with moisture ingress, especially if you use Scotchlok-type connectors. It's better to make the connections inside the car, under the dash. Select the correct wire by using the test lamp and switching the indicators on. The test lamp bulb should flash on and off in conjunction with the vehicle's indicator lights.

Top right. This sequence shows the fitting of a pin switch to be operated by the bonnet/hood; however, the method is the same for any point of entry. It is important to ensure that the switch will not get too wet. The bonnet should be able to close normally but still press the pin in far enough to operate the switch. This involves some careful measuring beforehand. Centre right. Drill a hole of suitable size and clean up the rough edges (burrs) with a round file. As the switch forms the earth (ground) connection through itself to the vehicle body, it has to make contact with bare metal. To prevent rust, use a copper-based grease to repel moisture but allow electrical conductivity. The use of plastics in this area of the engine bay is increasing and you should ensure that the switch is not mounted in plastic unless you are prepared to run an earthing wire to the switch body...

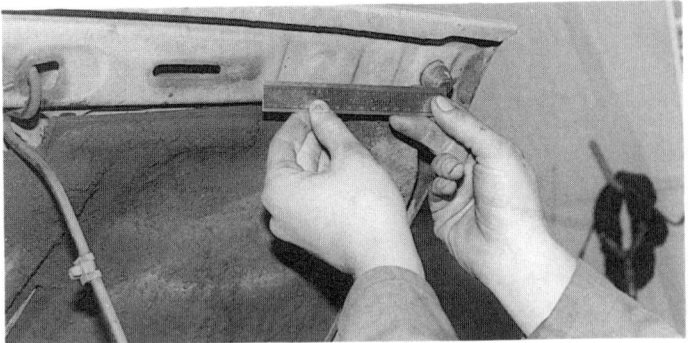

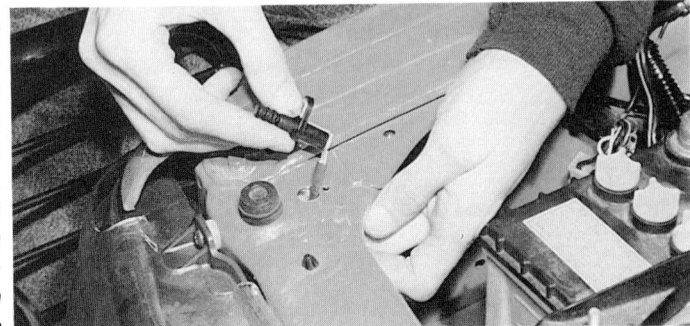

Above. ...The method of fixing pin-switches varies little; the switch is inserted through the hole ...

the circuit. In practice, it demands some skill and knowledge from the fitter, especially if such things as fuel injection are to be tampered with. CAUTION! For the do-it-yourselfer it is best to stick to immobilisation via the starter motor - if the alarm is trig-

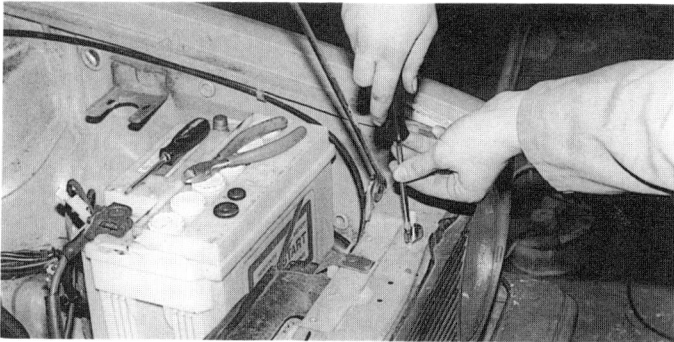

...and secured by a large nut or by a screw through its baseplate. The wire from the alarm is attached to the single terminal at the back of the switch. This connection must be secure and not fouling on bodywork or ancillaries.

gered, the starter motor will not operate. That way, faults occuring whilst the vehicle is on the move will have no effect. If you are in the slightest doubt, even if you install the rest of the system yourself, it's best to have the immobilisation aspect taken care of by a professional. If you do install the immobilisation system, please have it checked over by a professional before using the car.

❏ PIN-SWITCHES

Do not confuse with *voltage-drop* sensing, where the power taken from the courtesy light circuit is used to trigger the alarm.

As the name suggests, this method of alarming an entry point relies on a mechanical pin-switch, rather than complex electronic sensing: the alarm triggers as the door, etc., is opened. Where the doors and/or boot (trunk)/hatch (tailgate)/bonnet (hood) have courtesy light pin-switches, it is usually possible for them

to be pressed into service. If not, new pin-switches must be fitted. Where appropriate, your alarm will come with at least one pin-switch in the box. It is possible to buy extra switches as options.

Some owners don't bother with pin switches on the doors when some form of interior sensing is fitted. But wise owners do for, although it is truly a 'belt and braces' method, you cannot have too much security.

❏ INTERIOR SENSORS

Electronic alarm systems use ultrasonics, microwaves or other sensors to detect movement. However, all these sensors are fitted and set-up in much the same way. The alarm manufacturer's instruction manual will recommend the positioning. Invariably, there are two sensor heads, a transmitter and a receiver. The relevant beams are issued by one, collected by the other and the electronic circuitry remembers the way they were received at any given time. Should that pattern vary (say, because a hand comes through a window) the alarm will sound.

There are two basic methods of mounting the sensor heads. If they are separate units, they can be mounted at the top or bottom of the vehicle A-post (each side of the windscreen/windshield). Care should be taken not be trap the wires. It's usually possible to prise the trim away from the A-posts, run the wire up inside and then refit the trim.

53

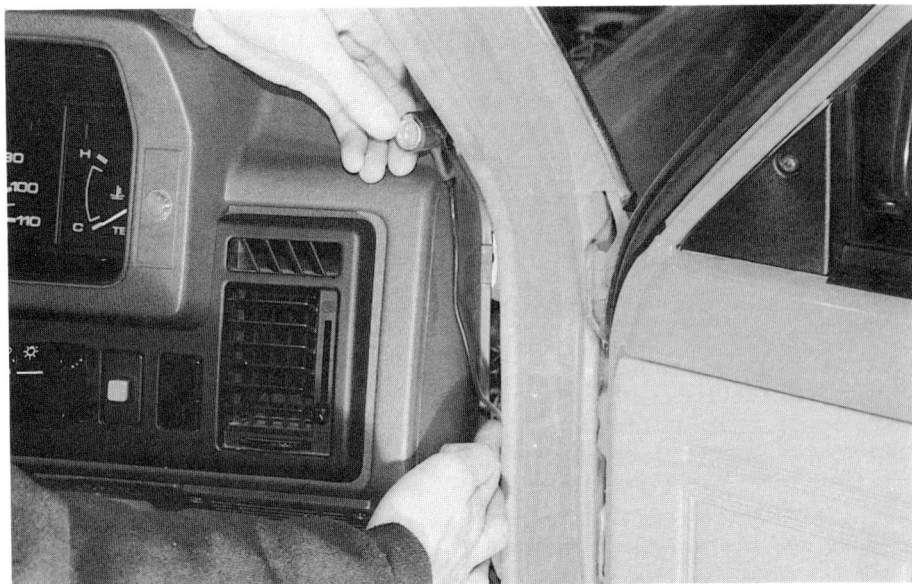

Once you've decided where on the A-post to position the sensor it's usually quite easy to hide the wires behind the post's trim.

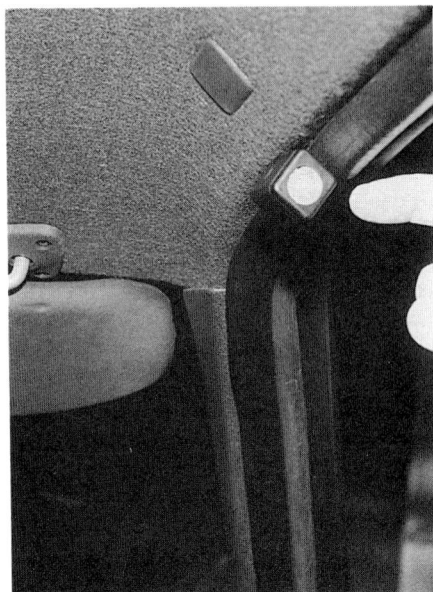

Sensor mounted well clear of the sun visor and wires neatly hidden behind the door trim strip.

Pod-mounted sensors can be positioned under the dash. Pods often have a sticky fixing pad.

When mounting the sensors at the top of the pillar, ensure that they will not be blocked by the sun visor. The sensor control unit is usually mounted under the dash area, (unless it is part of a central control unit positioned under the bonnet/hood). Remember that wherever a wire passes through a hole in metal, a grommet MUST be used.

Where the sensors are enclosed in a single pod, it can be mounted on the rear shelf or, more commonly on the dashpanel. Better still, from an aesthetic point of view, is to position the unit under the dash. Note that a double-sided sticky pad is sufficient to secure the light-weight unit (thus saving damaging the under dash by drilling holes) and that the sensors are connected to the alarm by means of a simple plug.

On no account should sensor wires be cut. If they need extending, then a proprietary extension should be used. If there is too much wire, it should be coiled neatly, taped and secured out of harm's way using electrician's tape.

During the setting-up procedure, you may have to adjust the sensor sensitivity, so it's best not to completely refit the trim until you're sure everything is absolutely right.

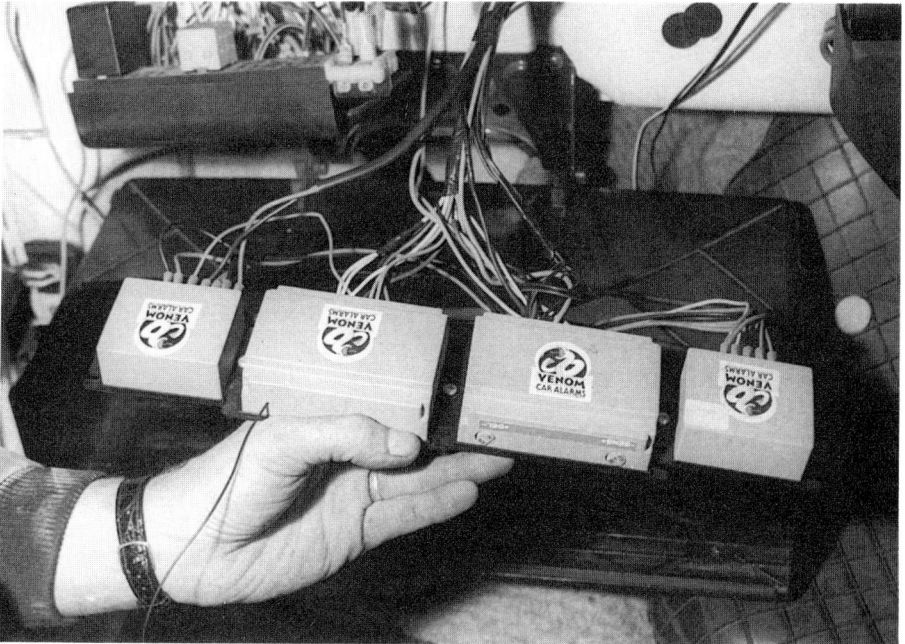

More complex systems are likely to have a number of interface units which, ideally, should be in the car's cabin. Here, no fewer than four interfaces are secured quite handily on the top of the glovebox. It's often possible to secure interfaces using velcro pads or double-sided sticky pads. Don't refit all that trim until you've tried the alarm to make sure it works - there's an awful lot of work in removing it again!

❏ MOUNTING INTERIOR ELECTRONICS

Depending on the type of alarm you fit, you may have to fit the electronic 'brains' inside the car. Certainly, if you add other functions (self-closing electric windows, etc.) to your system, you will have electronic interfaces to find homes for. The underdash area is a favoured position, where it's warm, dry and safe.

❏ SETTING UP

Having physically installed the alarm system, many owners assume that the job is finished. It isn't, for it must now be adjusted to suit the vehicle and the owners' particular requirements. CAUTION! If your alarm system includes any form of engine immobilisation (apart from a starter motor cut-out) then it is vital that the system CANNOT be armed whilst the ignition is on. So that is check number one - switch on the ignition and try to arm the system. If it won't arm, then you can carry on through the rest of the suggested setting-up procedures. If it WILL arm, then you MUST find the source of the problem before going any further. To run the risk of the system arming (and cutting-out the engine) whilst the vehicle is in motion is extremely dangerous and unnecessary. If you cannot find a fault, then disconnect the alarm altogether and take it to a professional fitter for his verdict. It could, of course, be that the equipment itself is faulty.

Radio remote control range can be varied by moving the position of the aerial in the car. Interior sensors will need to be adjusted so that they trigger the alarm during a break-in, not when a car drives by. Shock/tilt sensors, too, need to be adjusted correctly.

All adjustments should be made in accordance with the alarm system manufacturer's recommendations. Some systems have a 'mute' facility on the siren, which means that your neighbours have a more peaceful time during testing. If your system hasn't, it's a good idea to drive to somewhere a little more isolated.

❏ FAULT-FINDING (ELECTRONIC SECURITY SYSTEMS) - NEW SYSTEMS

Electronic alarm systems are complex items - even the simplest of them - and so offering a specific fault-finding chapter covering all eventualities is like asking your doctor to diagnose your illness over the telephone. However, these basic guidelines should help if you have problems with the security system you have fitted.

Initial difficulties almost all come under three headings -

a) not reading the 'instructions for use' properly

b) faulty installation techniques

c) a lack of precision in installation.

With this in mind, problems encountered during the setting-up procedure should be assumed, initially at least, to be the fault of the installer. Your work should be checked accordingly against the manufacturer's fitting instructions.

Of course, no manufacturer is perfect and if, after checking through everything with a fine-tooth comb, you find that all is still not as it should be, then it could be the product itself at fault, in which case the retailer should exchange it. The adjacent chart covers common problems you're likely to experience.

FAULT DIAGNOSIS CHART

✗ ALARM DOESN'T ARM WHEN BUTTON PRESSED

☛ Use a test light to check power and earth (ground) wiring. Check all doors, bonnet (hood), etc., are closed. Check remote control unit battery has power (if applicable). Check alarm aerial position (if applicable). On systems with battery back-up, check override key is in 'on' position. If system is infra red, aim nearer the sensor and get closer.

✗ ALARM DOES NOT SOUND WHEN BONNET (HOOD)/BOOT (TRUNK)/ HATCH (TAILGATE) DOOR OPENED

☛ Has the alarm finished setting-up? Most require around 40 seconds to fully arm. Is there a re-entry delay? Check pin switches - they may be sticking. Check wiring to each pin switch (look for a break in the wire or a faulty terminal connection). Check that pin switch is earthing (grounding) correctly. Are interior sensors plugged in and working correctly?

✗ ALARM DOES NOT TRIGGER FROM INTERIOR SENSORS WHEN ARM INSERTED THROUGH WINDOW

☛ Sensors require sensitivity adjustment. Check that sensor plugs and sockets are the right way round and that they are secure.

✗ ALARM DOES NOT TRIGGER WHEN VEHICLE ROCKED OR PUSHED

☛ Shock sensor sensitivity set too low: adjust to suit.

✗ ALARM IS TRIGGERED BY SLIGHTEST TOUCH

☛ Shock sensor sensitivity set too high: adjust to suit.

✗ SENSORS TRIGGER THE ALARM FOR NO APPARENT REASON

☛ Ultrasonics are sensitive to wind pressure - is sunroof and/or window open? Check that interior air vents are closed. All sensors; sensitivity set too high? Check that there is nothing moving in the vehicle.

✗ INDICATORS DO NOT FLASH ON ARMING/DISARMING OR TRIGGERING AN ALARM

☛ Check alarm-to-indicator connections using test light. Check indicator bulbs and earth (ground). Note that not all electronic alarm systems are linked to the vehicle indicators.

✗ PASSIVE ARMING DOES NOT WORK

☛ Not all systems have passive arming. Check all wiring correct. Check all doors/windows/boot (trunk), etc., are closed. Have you allowed full arming time (usually around 60 seconds) after last door closing?

FAULT DIAGNOSIS CHART

✗ FLAT BATTERY

☛ Alarm systems drain the modern battery very little and should be able to be left armed for several days (at least) with no undue effects.

A flat battery problem could be the fault of the battery itself or current drain in any electrical system. Battery health can be checked by a battery specialist or by using one of several proprietary testers available. A test under load is important.

If the battery proves to be okay but still loses its charge easily, even after being fully charged, then you should attempt to isolate the cause of the drain. A multimeter makes this a relatively simple procedure, as it is merely a question of testing various circuits (this can usually be done at the fuse box) until the culprit shows up on the meter. If the alarm is found to be at fault. it should be returned to the shop/dealer from which it was purchased.

✗ CENTRAL LOCKING OPENS AS ALARM IS ARMED AND LOCKS AS ALARM IS DISARMED

☛ Polarity of C/L interface is incorrect. Note that not all alarm systems are connected to the car's central locking system.

✗ CENTRAL LOCKING (ELECTRIC WINDOWS/ELECTRIC SUNROOF) DOES NOT OPERATE IN CONJUNCTION WITH ALARM

☛ Check all connections for security and power using a test lamp. Make sure the individual features are working as normal. Is interior trim fouling on anything (eg, a trim panel on a central locking motor)? If power is present at interface but not at accessory, and all connections are sound, interface is suspect.

❏ FAULT-FINDING (ELECTRONIC SECURITY SYSTEMS) - EXISTING SYSTEMS

It may be that you bought a car with a security system already installed or, perhaps, you fitted a system some time ago and now find that all is not well. The fault-finding techniques detailed in the previous section still apply; however, you will have an extra problem if you do not have the system manufacturer's installation and/or setting-up instructions.

For example, you might think that the system is faulty because it doesn't 'chirp' in confirmation whenever it is armed or disarmed. You might be right, but some systems don't have this feature and on others it is switchable. Without knowing what it SHOULD be doing, you can't tell when it's not doing it!

In this situation, your first port of call has to be the alarm manufacturer to see if you can obtain a handbook/ fitting instructions. If the system is 'factory fit' then the car's manufacturer or its local dealers should be able to help.

If you know that the system was professionally fitted at a specialist dealership, your enquiry would be better made direct to them rather than the system's manufacturer.

Whoever you approach, you should remember that you'll need to prove that you are the legal owner of the car and that you have the alarm system keys.

PROTECTING AUDIO EQUIPMENT

HOW TO PROTECT RADIOS, TAPE AND CD PLAYERS

Some thieves wouldn't dream of taking your car - the hassle of trying to sell something that big is far too much like hard work (with not a little possibility of being caught). But your bright, shiny stereo gear - that's much more like it! In general, it's simple to remove, simple to hide and simple to sell; there's an ever-growing market for 'slightly used' in-car hi-fi equipment, no questions asked. So, even though your car itself may not be high on the thief's 'wanted list', there may be something inside it that is ...

There are many ways to protect your precious noise-makers, as shown later, but the same basic rules of survival remain the same; park sensibly, lock the doors/hatch/windows, don't leave valuable items on show etc. It all helps to make life more difficult for the thief. If you've got some really interesting equipment fitted, you may be tempted to plaster your windows with stickers advertising your aural allegiance. Why not go the whole hog and stick a flag on the roof with the message "This car has lots of kit worth stealing inside"? Make the thief work for his illegal living - KEEP HIM GUESSING! But perhaps you're not really interested in ICE. OK, so the manufacturers supplied a rather average radio/cassette deck in the car, but you only use it occasionally to check the cricket scores. Well, *you* may not be interested, but the thief will be. You see, the market for ICE does not revolve around the top-end machines - just as car thieves don't just steal Ferraris. And the problem is that, whilst the stereo might not be worth much, he could well do £500 worth of damage to the dashboard and your paintwork getting it out - when the tools of his profession are building bricks and crowbars, that sort of thing tends to happen.

A lucky escape for someone - in this case the thief didn't destroy half the dashboard as he took the radio/cassette unit! ICE (in-car-entertainment) equipment is in the frontline of the security war and it's up to you to make sure your radio/tape player/CD unit is not a casualty.

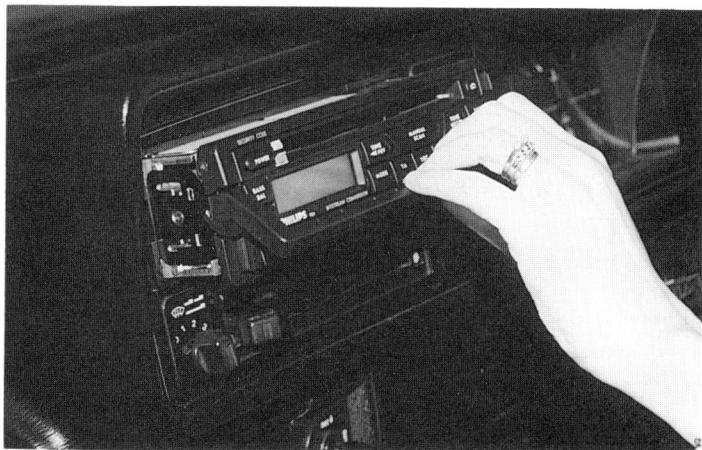

A removable audio unit is a wise choice and most manufacturers now offer them. (Courtesy Philips Car Systems)

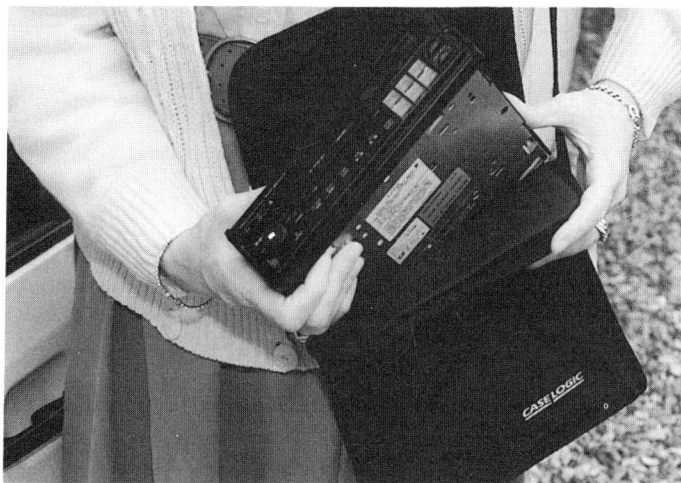

A removable set can be quite cumbersome, so a specially designed carrying case such as this can be useful. (Courtesy Caselogic)

❏ REMOVABLE SETS/REMOVABLE CONTROL PANELS

The best way to ensure that your car's audio equipment is not removed illegally is to remove it yourself. Most manufacturers produce at least some removable decks, whereby it is a special frame, not the set itself, which is actually fixed and wired into the vehicle. Many folk (unwisely) leave the set in the frame or (not so bad) lock it in the boot because carrying the deck around is somewhat awkward. One answer is provided by Caselogic, who produce a handy and stylish carry case designed to suit all DIN size decks (including CD players). It's also got 'pockets' for cassettes and/or ancillaries.

Taking the same idea a stage further are units with removable control panels or part control panels. Of course, these are much easier to carry around than the whole set. Some of these sets feature an LED

Under the removable front panel of this Roadstar set is a flashing LED light to warn off thieves. If the set is stolen an internal siren sounds for up to 30 minutes!

This kind of safe, which is designed to be fixed in the car's luggage compartment, is a practical alternative to carrying audio equipment with you. (Courtesy Clarion Shoji (UK) Ltd)

which flashes when the panel has been removed.

As an example of what car manufacturers are doing to improve ICE security, the Peugeot 106 features a radio/cassette system likely to become increasingly popular. It comes in two parts, the readout and the control panel. Better still, it will only fit this particular car, making it useless as an item for general sale - legal or otherwise.

Roadstar broke new ground with a range of radio/cassette decks featuring their SKIP

Blaupunkt's 'Keycard' security system. (Courtesy Robert Bosch Ltd)

system. Under the removable front panel is an LED light which flashes a warning to the thief. However, if the thief ignores this and takes the deck a 110 dB siren, built into the deck, will sound - and keep on sounding for up to 30 minutes! The siren is powered by its internal alkaline batteries and Roadstar are developing a unit with batteries rechargeable by a trickle charge from the vehicle charging system.

If you have a removable deck but can't (or don't want to) take it with you, the Audio Safe is a strong, steel box which offers far greater protection than a glovebox. It is fitted with a radial pin lock and, being permanently mounted in the luggage compartment of the car, presents even a determined thief with far more hassle then he wants.

❏ **CODED SETS**

An electronically coded unit is also a good idea. There are upwards of 10,000 different combinations for a typical 4-figure code, so there's little danger of it being broken. The latest decks have their codes burnt into the 'chip', meaning that major surgery is required in order to produce a working set. The 'coded set' stickers provided should be prominently displayed in the front windows, in order to prevent the thief from breaking in to find out what's inside.

Blaupunkt's Keycard system mixes security coding with a credit card sized programme card. Without the Keycard, the set is useless and of no value to the thief and the card actually records all the settings of the deck (display colour, loudness, etc.). The company is so confident of its effectiveness that they operate a free replacement warranty, should the deck be stolen.

❏ **PHYSICAL SECURITY DEVICES**

The Metro radio/cassette lock is a cassette shaped device on the end of a piece of plastic-covered, braided steel cable. It locks into the cassette aperture and around the steering wheel - the radial-pin lock has 2,500 combinations. Even if the deck is removed, it can't be taken from the car without seriously dam-

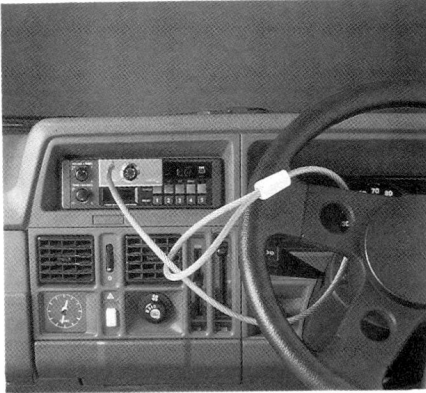

This product is bound to be a strong deterent to would-be ICE thieves. (Courtesy Metro Products Limited)

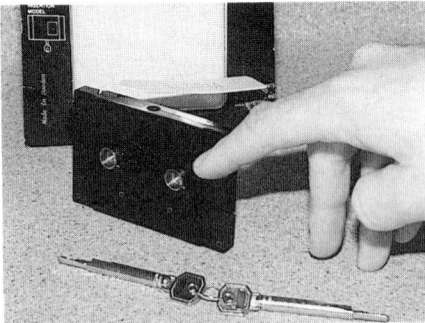

Motus device locks into cassette slot.

Locking devices like AudioLOK make ICE equipment much more secure.

aging the set. As ever, it has a high visibility factor.

The Motus Security Lock is also shaped like a cassette, but this time is locked into the deck by using a special key. A rude, yellow tongue, embossed with the word 'alarm' in red, protrudes from the end to warn thieves to look elsewhere. Removing the device means wrecking the deck.

The Quickfit 70 'AudioLOK' is another variation on a theme. It requires the removal of the deck and cage (a once-only operation) for the fitting of a special security bracket. Once in place, a strong steel cover can be locked into place, covering the whole deck. It is secured by a radial-pin tumbler lock and its bright yellow lettering ensures high visibility.

If your set (or bits of it) is not removable, then covering it up prevents the thief from seeing what is there. Devices like the 'Hideaway' simply clip into position and, at a glance, look like the blanking plate manufacturers use when no audio equipment is fitted.

❏ **TAPE CASSETTE/CD SECURITY**
Your collection of cassettes or CDs is of great interest to the thief. You should always keep cassettes so that the spools cannot move (either in their cases or in purpose-made boxes). You can purchase specially designed storage boxes which will conceal your tapes or CDs.

❏ **AERIALS/ANTENNA**
It's unlikely a thief would want the hassle of relieving you of your aerial/antenna, but it draws the vandal like a magnet if left extended. It doesn't take long to press the aerial down

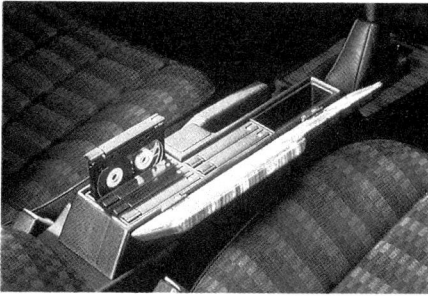

when you leave the car. The best bet from a security point of view is an electrically-operated aerial/antenna which retracts when the audio unit is switched off.

Many of the roof mounted units (and cellular 'phone aerials) can be unscrewed and left in the vehicle - again, it doesn't take long.

IMMOBILISERS
DEVICES THAT STOP YOUR CAR FROM BEING DRIVEN AWAY

Interrupting vital systems on a vehicle to prevent its operation is hardly a new idea (even though car crime is now at epidemic levels), and has been around since the first car sputtered into life. Careful owners would get into the habit of removing the distributor lead (or sometimes the distributor cap) whenever their car was parked in public. A variation on this was the cut-out switch, a simple on/off device, usually placed in line with the ignition coil. By definition, the switch would have to be user friendly and easy for the driver to use, which usually meant easy for the thief to use, too!. Times change, and modern electronic/hydraulic immobilisers use complex modern technology to give the thief a hard time.

Immobilising your car by means of an electronic or hydraulic immobiliser is not only a good idea, it is an essential requirement with some insurance companies before they will insure certain vehicles. However, it CANNOT be recommended that fitting of dedicated vehicle immobilisation devices be tackled by the vehicle owner. Apart from the safety connotations, there is also the insurance angle - some companies offer discounts, but only where the device has been fitted professionally.

There are a number of purpose-made immobilisers on the market. By definition, they are not full electronic security systems; they do not make a siren sound nor flash the vehicle's lights.

They are usually controlled by an electro/mechanical key, though some can be wired to work passively (to arm automatically at a pre-set time after the last door has shut) and some can be linked to an electronic alarm and switched via the remote key.

Their aim is to disable your car so completely that it cannot possibly be driven away, regardless of how clever the thief is.

Caution should be used when selecting and fitting any immobiliser, the counsel here relating to what happens if something goes wrong. It was certainly the case that some early immobilisation systems could land you in serious hot water if they malfunctioned: if a system designed to cut the fuel or engine electrics went wrong at speed on a motorway, for example, the results don't bear thinking about.

However, the progress of microprocessor technology has meant that the likelihood of anything untoward happening is now much reduced and, in some cases, eliminated altogether. Nevertheless, once you have ascertained from the seller what the immobiliser can do in terms of securing your car, your next question should be "what happens if it goes wrong whilst the car is moving?" What you are looking for is a device which fails safe.

In terms of warranty, many of the top companies are sufficiently confident in their products to offer quite a comprehensive back-up, including mobile breakdown cover if you hit problems.

The 'Vecta' immobilisation system is the one which first brought this kind of security device to the fore, not least because of the link between it and insurance discounts. At the time of writing,

Two versions of the UniValve, a device which locks the vehicle's brakes. (Courtesy Redsure)

more than ten insurers specify, recommend and/or endorse the product. It works by controlling up to 8 systems and/or sensors and is operated simply by removing the special key from a socket in the dashboard/console etc. It MUST be professionally fitted and cannot be obtained other than through approved dealerships.

The 'UniValve' is a modern development of an American device (called the 'Autosafe'). It is operated by a Swiss-designed key with 1.5 million permutations and no blanks or masters. To activate the system, the key is turned and the car's brake pedal pressed. This locks the brakes in the 'on' position. (If the owner forgets to press the brake pedal, the system will lock as soon as the thief presses it.) In addition, the system breaks the starter motor circuit, which creates another problem for our thief. A flashing LED mounted on the dash

warns the thief of the system's presence. It is a product for professional fit only and UniValve will only sell their product to accredited dealers. The warranty will only be honoured if they have received a fully-completed report form stating that the device has been professionally fitted to a specific vehicle. According to the distributors, UniValve can be fitted to cars (including 4WD), motorcycles, and commercial vehicles.

Laserline's 994 is a passive immobilisation system which activates automatically twenty seconds after turning off the ignition. Passive arming is becoming *de rigueur* when such electronic devices are fitted, especially as far as insurance companies are concerned. The 994 can cut up to four different circuits and a flashing LED warns the thief of the system's presence. The car's owner disarms the system by means of an electroni-

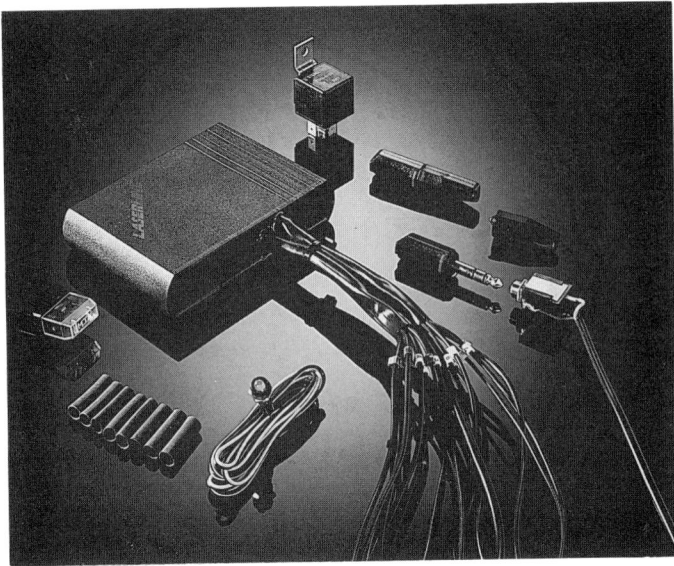

Laserline 994 is an electronic system which immobilises the car by cutting several essential electrical circuits. (Courtesy Laserline Car Alarms)

disarm the unit, the code is changed. This random changing is based on a 96 bit binary code stored on the decoder key, the permutations of which are impressive, to say the least. According to the company, if you started suggesting random codes 50 billion years ago, there is still only a 1 in 1,000 chance that you could hit the right code today!

The immobilisers shown and described in this chapter are but a handful of those available on the market and, as with most things, the best advice is to shop around. Take independent advice and seek out unbiased test reports. Your local crime prevention office may have had experience of vehicle immobilisers, so it's worth giving them a call. Don't forget to balance a cost differential against a possible insurance discount. But beware, whilst most immobilisation products are high quality items and safe to use, some are not so hot.

cally-coded key and even the key has security features built in. If required the system can be linked to a siren.

Kemitron Automotive is an electronics company which, amongst others, has produced the dashboard for the Aston Martin Lagonda and the Jaguar XJ220. Their experience has been used to produce an immobiliser using BS standard components. The KA44 immobiliser is a completely sealed unit which offers the facility to disable up to four circuits on the vehicle. One of these is permanently active whilst the other three are randomly and constantly changed. The device is activated (only with the ignition off) by pressing the decoder key against the dashboard mounted 'port' - the latter being a small item having the appearance of a watch battery. Each time the key is used to

❏ CHOOSING AN IMMOBILISER
Here are the important points to consider before making a choice of immobiliser -

✓ Does your car warrant an electronic/hydraulic immobiliser? Owners of high-risk cars (classic, sports, high performance hatchbacks, etc.,) will benefit most.

✓ Almost invariably, an immobiliser will have to be professionally fitted. Build this into your budget.

✓ Don't buy an immobiliser second-hand.

✓ If you already have an alarm system, some immobilisers can be linked into it and operated via the remote handset. Check before you buy.

✓ Some immobilisers are passive, *ie*, they operate automatically whenever the car is left. Do you want this? Is there the option for manual control?

✓ Is there an insurance discount available? If so, it effectively makes the device cheaper. Check out the real cost and make sure you still get what you want from the policy (in terms of cover, bonus protection, excesses etc.)

✓ Ask the question "What happens if the device develops a fault whilst the vehicle is moving?" You need it to fail safe!

✓ Don't be complacent. An immobiliser won't stop a thief damaging your car or taking the contents. It's best to fit an electronic alarm as well.

✓ Warranty. Make sure you get a good one, there's no reason not to. Some companies offer a mobile breakdown service for system-related faults.

✓ Some manufacturers offer a check-up service for their products. It's worth taking-up, as nothing lasts forever, and suspect parts can be replaced before they go wrong.

WHEN THE WORST HAPPENS
WHAT TO DO IF <u>YOUR</u> CAR IS STOLEN OR BROKEN INTO

❏ WHAT TO DO IF YOUR CAR IS STOLEN

1) "Don't panic" is easier said than done, but it's vital to remain calm. Make sure that your car really isn't where you left it, particularly if you used a multi-storey car park. All floors tend to look alike, and there are many cases of owners reporting their cars missing only for them to be found some days later by car park officials. Apart from your red face and inconvenience, you'll have wasted a lot of Police time and money.

2) If you car *has* disappeared, then contact the Police immediately. They'll need to know all the details of the car so it's best to take a couple of minutes before your call just to jot them down; registration number, colour, make and model and any distinguishing features, such as alloy wheels, stripes etc.

3) You'll also have to list what was in the car. Not as simple as it sounds. Try to remember how much fuel you left in the vehicle - this will give the Police some idea of the range of the car - thieves seldom stop and fill the tank. If you've left your vehicle documentation in the car, it opens other avenues for the Police to pursue, as the thieves could well try and sell the car 'legitimately' - that's why you should leave such paperwork in a secure place at home.

4) Did you leave any credit/cheque guarantee cards in the car? If so, it's definitely in your interest to contact the banks or credit card companies involved, so that the thief cannot use them.

5) Did you leave your house keys in the car? Is your home address in the car as well? Tell the Police if this is the case, and they will be able to send someone to your house to check it's still secure. It's also a good idea to 'phone a neighbour and ask them to keep a watch over the house until you return: don't forget that the thief will be rolling up in your car and so, from a distance, everything will appear normal.

6) Whatever happens, if you've lost your house keys in this manner, you'd be wise to change the locks. Even if the keys are recovered with the vehicle, there is the possibility that duplicates have already been made. Having set the wheels of law in motion, it's a good idea to start letting your friends and family know what's going on, not least because you'll probably need some help to get home! Your insurers need to know what's happening and you'll have to fill in a claim form. It's easy to ruin your chance of compensation by filling this in incorrectly, so liaise with your insurers/brokers as to the correct wording.

7) If you've provision in your policy for a hire car, check if you have to hire from a particular company and/or if there is a daily limit. Some policies allow for the expenses of a night away from home if your car is stolen,

70

which could ease your problems considerably.

❏ WHAT TO DO IF YOUR CAR IS BROKEN INTO

1) If your car has not been taken but has been broken into, the previous comments apply. However, before you consider driving your car anywhere, you should ensure that it is legal and safe. As we know, thieves don't bother too much about damaging your car in pursuit of their ill-gotten gains. It could well be that vital wiring and/or safety components have been damaged. It's not a bad idea to ask the officer to whom you report the theft to have a glance over the car for you.

2) If you left your cellular 'phone in the car, you should get in touch with your air-time suppliers immediately. Once they know that the car has been stolen, they can put an electronic bar on all calls from your unit. It's worth checking at this point whether the thief has made any calls - most can't resist it. As mobile 'phone numbers are logged automatically it may give a clue to the thief's identity.

3) Whether the damage caused to your car is a case for your insurers will depend on the type of cover you have and the extent of the damage. Most comprehensive policies cover all glass in the same way as the windscreen in that it will be replaced without affecting your no-claims discount, but there is usually an excess. (Some companies will cover glass on third party or third party, fire and theft policies in return for an extra premium).

4) Probably the first thing you'll do after having your car stolen is start to think about vehicle security. You can console yourself that a high proportion of cars are alarmed only after they have been attacked and, whilst prevention is better than cure, it is also true that it's better late than never.

5) Certainly, you should never assume that because your car was broken into last night, it won't be broken into tonight as well. Complacency is the car thief's main ally.

❏ CHECKLIST

✓ Is your car really missing? Multistorey and other large car parks in particular are legend for producing false alarms as owners end up on the wrong floor or area and panic.

✓ What was inside the car? The police and your insurers will need to know. Credit cards, cheque books, keys?

✓ Contact the police. They need to know everything about the car and its contents. Take the name and number of the officer you report the theft to for future reference.

✓ If you left credit cards or cheque books in your car, contact the relevant banks/companies immediately to reduce the risk of fraudulent use.

✓ Contact your insurers. They'll need to know your policy number, everything you told the police, plus the name of the station/office you reported it to. If you have a hire car provision in your policy, check if there is a cash limit per day. Also, do you have to gain their approval or hire from a specific company?

✓ Was your car fitted with a security device which included a theft warranty? Some devices promise a cash payment if a thief takes your car when so fitted.

✓ Cellular 'phone. If you have a carphone, inform the air-time suppliers quickly, so that they can bar calls from the unit. Check also if any calls have already been made - it could give a clue to the identity of the thief.

GLOSSARY OF TERMS
WHAT ALL THOSE TECHNICAL TERMS REALLY MEAN

❏ AUTO RESET
Having sounded for around 30 seconds, alarms have to reset. This has the dual effect of keeping the neighbours happy and not flattening your battery. Were this not so, a clever thief could deliberately trigger the alarm and wait until the battery drained before raiding the car.

❏ BATTERY BACK-UP
Highly recommended. The alarm includes a nicad battery which usually takes a trickle charge from the vehicle battery. If the power supply to the alarm is cut for any reason, the alarm will sound immediately and then continue to function as normal.

❏ CODE
Many in-car audio devices (radio/cassette/CD) are equipped with a security code which has to be entered if the power to the set is switched off. With more than 10,000 permutations for a typical 4-figure code, such a set is recommended.

❏ COMPACT ALARM (See also Modular Alarm)
An electronic alarm system where the electronic control components are contained within the siren housing and mounted (usually) in the engine bay.

❏ CURRENT SENSING
This is a feature known also as voltage-drop or current drain. The alarm system monitors the current drain from the battery and if it changes (say, because an interior light comes on as a door is opened) then the alarm sounds. Where an electric cooling fan is fitted to the engine (which can operate after the engine has been switched off), a special by-pass circuit is necessary to prevent false alarms. Some systems offer this, others don't. Current sensing can sometimes be fooled during cold weather when the battery voltage level drops substantially during the night - hence those 3am 'alarm' calls! Some of the latest alarm systems feature 'intelligent' sensing, whereby the microprocessor can tell the difference between, say, a cellular 'phone operating after the ignition has been switched off and an attempted break-in.

❏ DELAY
An alarm feature whereby the means of arming and disarming the system is actually inside the car. This means that there has to be a delay between opening the door and switching off the alarm.

❏ DIRECT CONTACT
Also known as earth (ground) contact, this circuit utilises mechanical pin-switches on opening points (doors, hatch/boot (trunk), bonnet (hood)). These could be the original switches or accessories provided with the alarm.

73

❏ EXTERNAL KEY SWITCH

Once used extensively, now quite rare and never found on the more expensive systems. Though it is an obvious and visible deterrent, the key switch is also unreliable and (relatively) simple to disable. Choose an electronic remote control device every time.

❏ GRABBERS (See also Scanners)

A device used by the thief to ascertain the electronic code used by your alarm system. As you arm your system, the Grabber electronically 'listens' to the code reference and stores it for the thief to use. The thief has to be fairly close to your car for it to work and many alarms now have constantly changing code patterns to help defeat this device.

❏ IMMOBILISER

A device to prevent the car being driven away. Many electronic alarm systems include immobilisation, though there are varying methods of preventing the car starting. Some disable the fuel pump or interrupt the electrical feed to the ignition system. Immobilising the starter motor circuit is generally held to be the safest method of immobilisation, certainly for the amateur fitter. Some alarm systems offer the facility to immobilise more than one circuit. There are immobilisers on the market which are separate from any alarm system, some of which can be linked (interfaced) with a standard alarm system. There are also immobilisers which work directly on the vehicle hydraulic and fuel systems.

❏ INTERFACE

Any alarm interface is a device which allows a separate piece of electronic gadgetry to work in harmony with the alarm itself. With good alarm systems there is the option to link the central locking and/or electric sunroof/electric windows to the alarm. This means that arming the system also closes all points of entry automatically. It is common for top-of-the-range alarms to have a built-in interface for central locking, which saves having to wire and mount a separate unit. It is recommended that you link as many options as you can to the alarm system. The easier it is to secure your car, the more likely you are to do it - every time!

❏ INTERIOR PROTECTION

Once the sole province of the ultra-sonic sensor, many variations on the theme have appeared. Microwave sensors (see later) are used by some companies, whereas others have gone a different route. Philips, for example, have TRI-Sonic sensors which are reckoned to be so stable they can be used in a convertible. Digital Alarms also claim great things for their Microsonic sensors.

❏ LED

An acronym for Light Emitting Diode. Most alarms feature at least one of these which is positioned in a prominent place (usually on the dashboard) and which flashes to show that the alarm is armed. Expensive systems often have more than one LED which flash to show the nature of alarm violations.

❏ LOCKING WHEEL BOLTS/NUTS

Purchased in sets of four, these are designed to suit specific vehicles. One standard bolt/nut per wheel is replaced by a locking version. It is tightened using the standard socket and the specially shaped key. Without the key standard sockets cannot get a grip on the bolthead/nut.

❏ MICROWAVE SENSOR

Not a device for frying the thief, but a rival to the ubiquitous ultrasonic sensor. Microwaves can be transmitted through various materials, including vehicle trim and glass and, unlike ultrasonic waves, are unaffected by air movement. This makes them particularly useful in open-top vehicles.

❏ MODULAR ALARM (See also Compact Alarm)

An electronic alarm system where the electronic control system and the siren are mounted separately. It is usual for the siren to be positioned in the engine bay and the electronics to be sited under the dash in the cabin. Some alarm systems are particularly versatile and can be fitted in either modular or compact mode.

❏ PAGING ALARMS

A transmitter/receiver system using a miniature pager which bleeps and/or flashes a warning that the vehicle is being tampered with. The range of these items varies: as far as 2 miles (3.2 kilometers) is often quoted in the open, though this shrinks to around 220 yards (200 metres) in town. All pagers require a certificate of type approval to be legal in the UK and the rules are likely to be similar in other countries. Check before you buy.

❏ PANIC BUTTON

A feature operated from the alarm remote control handset. When operated, the siren will sound for a pre-set time and, if the alarm is so equipped, the indicators will flash. It is designed as a personal attack feature and will work regardless of whether or not the alarm is set.

❏ PASSIVE ARMING

A feature which sets the alarm without manual instruction to do so. Most set the alarm 30/45 seconds after the last door is closed (utilising the original courtesy light switch). More complex systems rely on a signal received from the remote handset; as you walk away with handset, the signal gets weaker and when it cannot be detected at all, the alarm sets. For those of a forgetful nature, or if the car is being used by a number of different drivers, passive arming is a good idea. This feature can be found on some immobilisers as well as full alarm systems.

❏ REMOTE CONTROL HANDSET

As it suggests, a small handset used to arm and disarm the alarm system. There are two types: radio remote control which MUST be DTI approved in the UK (the handset will have an MPT number on it) and infra-red. Radio remote control generally offers a longer operating distance than infra-red, which also tends to be adversely affected by reflections of sunlight on the vehicle glass. Some alarm manufacturers have several buttons on the handsets, allowing them greater flexibility to control ancillary items (such as electric windows) and even automatic garage doors.

❏ SCANNERS (See also Grabbers)

An electronic device which works in a random manner, electronically 'suggesting' a myriad of different code permutations to your alarm system. When it hits the right combination (rather like listening to the 'tumblers' on a safe) it enables the thief to switch your alarm off and remove your car or its contents. Many systems are now anti-scannable or, at least, incredibly difficult to scan.

❏ SHOCK SENSOR

Early shock (or impact) sensors gained a reputation for unreliability. This was partly due to their construction as simple pendulum devices, and partly due to owners setting them to be too sensitive. Modern shock sensors are electronic and much more controllable. Their aim is to trigger the alarm when a sudden shock is applied to the vehicle, such as the breaking of a window or violence to another point of entry. Companies such as Clifford have a variety of sensors which can be specifically tuned to detect the breaking of glass, etc.

❏ SIREN

The bit of the alarm that makes the noise! Some early alarms utilised the vehicle horn, though not to much effect as it could easily be disconnected. Also, modern sirens are much louder, especially if fitted in the vehicle cabin. Around 120 dB is the norm. Some alarms have a facility to run one or more (slave) sirens. There is some debate about how powerful they should be with regard to noise pollution, etc. As such, you should make sure that you are totally famil-

iar with the latest regulations when you buy.

❏ TILT SENSOR

This sensor measures the angle of the vehicle at the time the alarm was set. If this varies by a few degrees either way (the sensitivity varies) then the alarm is triggered. The tilt sensor is primarily designed to detect the wheels being stolen and/or the vehicle being loaded up onto a trailer.

❏ ULTRASONIC SENSORS

These are twin units (1 transmitter/1 receiver) which are mounted inside the vehicle. They transmit a beam of ultrasonic sound through the vehicle and if it is broken (say, by a hand coming through a window) the alarm is triggered. They can easily be upset by the sudden movement of air, which means that the sunroof/windows cannot be left open. Even a fresh air vent on a windy day will have to be closed. Some can be too sensitive to sudden changes in temperature and cause false alarms.

❏ VALET/SERVICE/OVERRIDE KEY

A key which overrides the alarm functions of systems with back-up batteries when the vehicle is in for service or cleaning. It can also be used if the remote key has been lost or has flat batteries.

❏ VOLTAGE-DROP SENSING

See current sensing.

MANUFACTURERS AND SUPPLIERS
USEFUL ADDRESSES & TELEPHONE NUMBERS

This list shows manufacturers and suppliers of security related products. The key below denotes the broad areas that each is involved in. Remember that addresses and telephone numbers change periodically so you may need to check with a commercial directory.

❏ **MAGAZINES**

Most motoring magazines touch on the subject of security at some point, but the following are likely to have more experience than most on the subject -

Which?
The Consumers' Association,
2 Marylebone Road,
London, NW1 4DF,
England.
☎ 071 486 5544

Car Stereo & Security,
DIR Publishing Ltd.,
PO Box 771,
Buckingham, MK18 4HH,
England.
☎ 0280 812197

Car Hi-Fi,
The Evro Publishing Co Ltd.,
60 Waldegrave Road,
Teddington,
Middx, TW11 8LG,
England.
☎ 081 943 5943

❏ **TRADE BODY (UK)**
C.R.I.S.P.
Communication House,
1 Kings Road,
Crowthorne,
Berks, RG11 7BG,
England.
☎ 0344 761272

❏ **MANUFACTURERS & SUPPLIERS**
AAA - Audio Analysis Alarms (Sonix)
☎ 0952 820284
Peartree Farmhouse,
Longford Road,
Newport,
Shrops. TF10 8LP,
England.
✶ - B

KEY TO COMPANY SPECIALTIES
✶

A - FITTED ELECTRONIC ALARM AND ASSOCIATED ACCESSORIES
B - ELECTRONIC ALARM DEVICE NOT PERMANENTLY FITTED
C - ELECTRONIC IMMOBILISERS (As part of a system or as stand-alone items)
D - SYNTHESISED VOICE ALARMS
E.- PHYSICAL LOCKING DEVICES (STEERING WHEEL/GEAR LEVER ETC.
F - RADIO/CASSETTE LOCKING DEVICES
G - PROTECTIVE GLASS
H - COSMETIC SECURITY PRODUCTS (RADIO/CASSETTE COVERS ETC.)
I - HYDRAULIC DEVICES (BRAKES/ CLUTCH)
J - MISCELLANEOUS (INC PAGERS ETC)
K - WINDOW ETCHING
L - I.C.E. SECURITY

ACTIVE '8'
☎ 061 653 5870
Bower Street,
Oldham,
Lancs, OL1 3PH,
England.
✶ - C

ASHLEY SECURITY
☎ 0922 409533
PO Box 166,
Willenhall,

W. Midlands, WV12 5QU,
England.
✶ - A, C

AUTOBLOCK
(See Tritec Distribution Ltd.)

AUTOCAR ELECTRICAL LTD
☎ 071 403 5959
77-85 Newington Causeway,
London. SE1 6BJ,
England.

✶ - A, C, K

AUTOEUROPE ELECTRICAL LTD
☎ 081 641 1999
Unit 6, Minden Road,
Kimpton Road Trading Estate,
Sutton,
Surrey, SM3 9PF,
England.
✶ - A, C

AUTO-LOC
See GMS Trading Ltd.

AUTOLOK (cassette lock)
(See Quickfit 70 Ltd.)

AUTOLOK SECURITY PRODUCTS LTD
☎ 0621 624 8171
Park Lane,
Royton,
Oldham,
Lancs. OL2 6PU,
England.
✶ - E

AUTOMANOR
☎ 0491 410329
4, Hernes Estate,
Henley on Thames,
Oxon. RG9 4NT,
England.
✶ - E

AUTOMAXI LTD
☎ 0525 383131
Chiltern Trading Estate,
Grovebury Road,
Leighton Buzzard,
Beds. LU7 8TU.,
England.
✶ - F

AUTOMOBILE IMPORTS LIMITED
☎ 0525 382713
7 Old Chapel Mews,
Lake Street,
Leighton Buzzard,
Beds, LU7 8RN,

England.
☆ - F

AUTO PRODUCTS INTERNA-
TIONAL LIMITED
☎ 0902 726598
Units 5-7, Planetary Road,
Wolverhampton, WV13 3XB,
England.
☆ - A

AUTOROOFS LTD
☎ 0384 636366
Maypole Fields,
Cradley,
Halesowen, B63 2QB,
England.
☆ - A, C

AUTOSEEKER
(See Hunter Instruments Ltd)

AUTOWATCH
(See Demon Products Ltd)

AUTOTRONIC
(See Autoroofs Ltd)

BLUE SHIELD ALARMS
(See Phoenix Audio)

BOSCH, ROBERT LTD
☎ 0895 834466
PO Box 98, Broadwater Park,
Denham,
Uxbridge. UB9 5HJ,
England.
☆ - A, C, F

CARFLOW PRODUCTS LTD
☎ 0525 383543
Leighton Road,
Leighton Buzzard,
Beds. LU7 7LA,
England.
☆ - E

CASELOGIC
(See Path Group)

CEL SALES LTD
☎ 021 585 6505
Unit 2, Block 6, Shenstone
Trading Estate,
Bromsgrove Road,
Halesowen,
W. Midlands. B63 3XB,
England.
☆ - A, C

CLARION SHOJI (UK) LTD
☎ 0793 870400
Unit 1, Marshall Road,
Hillmead,
Swindon,
Wilts. SN5 7DW,
England.
☆ - H, L

CLIFFORD ELECTRONICS INC
☎ 071 498 0200

21 Aberville Mews,
88 Clapham Park Road,
London. SW4 7BX,
England.
☆ - A, C

CLUB
(See Portland Marketing)

COBRA
☎ 0923 240525
Ital Audio Ltd.,
K & K House,
Station Approach,
Rickmansworth Road,
Watford,
Herts. WD1 7LR,
England.
☆ - A, C

CRIMEGUARD   SECURITY
PRODUCTS
☎ 0225 790730
17 Pegasus Way,
Bowerhill, Industrial Estate,
Melksham,
Wilts. SN12 6TR,
England.
☆ - A, C

CTR ELECTRONICS
☎ 0702 614224
Unit 1, Stanfield Road,
Southend-on-Sea, SS2 5DQ,
England.
☆ - C

DEMON PRODUCTS LIMITED
☎ 081 667 1300
266 Selsdon Road,
Croydon,
Surrey. CR2 7AA,
England.
☆ - A, C

DIGICOM
☎ 0602 228451
4 Faraday Building,
The Science Park,
Nottingham, NG7 2QP,
England.
☆ - J

DIGITAL VEHICLE SECURITY
SYSTEMS
☎ 0494 792499
Unit 1B, Saxeway Business
Centre,
Chartridge Lane,
Chesham,
Bucks. HP5 2SH,
England.
☆ - A, C, I

DIS-CAR-NECT
See Richbrook International
Limited.

DOME SILENT ALERT
(See Digicom)

EAGLE CLAW
(See Ring Automotive)

EEC Ltd.,
☎ 0946 830220
2 Moorview Close,
High Harrington,
Workington,
CA14 4NX,
England.
☆ - C

ELECTROSYSTEMS
☎ 0992 34428
19 Fairways,
New River Trading Estate,
Cheshunt,
Herts, EN8 0NL,
England.
☆ - A, C

ENIGMA
(See CTR Electronics)

FAIL PROOF
(See Mapleman (UK))

FANFARE ELECTRONICS LTD
☎ 0494 446555
Copyground Lane,
High Wycombe,
Bucks, HP12 3XD,
England.
☆ - A

FOXGUARD LTD
☎ 0278 427212
1 Sedgemont Industrial Park,
Bristol Road,
Bridgewater,
Somerset. TA6 4AR,
England.
☆ - A, C

FUEL EX
(See Widdowson Dalebrook
Engineers Ltd)

GMS TRADING LTD
☎ 0778 380324
8 Market Place,
Market Deeping,
Peterborough,
PE6 8EA,
England.
☆ - I

GT ALARMS UK LTD
☎ 0926 882382
GT House,
Berrington Road,
Syndenham Ind Est,
Leamington Road,
Leamington Spa,
CV31 1NB,
England.
☆ - A, C

GAMMA ALARMS
Tel: 0462 670550
☆ - A, C, I

GEMINI ELETTRONICA
☎ 0905 756900
Wainwright Road,
Shire Business Park,
Worcester,
WR4 9FA,
England.
☆ - A, C

GUNSON LTD
☎ 081 555 7421
Pudding Mill Lane,
Stratford,
London. E15 2PJ,
England.
☆ - J +(Multi-meters and di-
agnostic test equipment)

HELLA LTD
☎ 0295 272233
Wildmere Road Estate,
Banbury,
Oxon.
OX16 7JU,
England.
☆ - A, C

HUNTER INSTRUMENTS LTD
☎ 091 491 1091
Unit 17, Octavian Way,
Teme Valley Trading Estate,
Gateshead,
NE11 0HZ,
England.
☆ - C

IMMOBILISER LTD
☎ 0245 350161
101 New London Road,
Chelmsford,
Essex. CM2 0PP,
England.
☆ - C

INVISIBEAM UK LTD
☎ 061 224 0769
Stanhope Street,
Levenshulme,
Manchester.
M19 3QA,
England.
☆ - A, C, D

KEMITRON AUTOMOTIVE LIM-
ITED
☎ 0244 537900
Hawarden Industrial Park,
Manor Lane,
Deeside,
Clwyd, CH5 3PP,
England.
☆ - A, C

KENWOOD
☎ 0923 816444
Trio Kenwood UK Ltd.,
Kenwood House,
Dwight Road,
Watford,
Herts. WD1 8EB,
England.
☆ - A, C

KEYSTOP ALARM SYSTEMS
(See AutoEurope Electrical
Ltd)

KINGAVON LTD
☎ 0473 219131
Parkside, Duke Street,
Ipswich, Suffolk, IP3 0AF,
England.
☆ - A, C, E

KLAMP-IT (INNOVATIONS) LTD
☎ 0604 721560
Unit 44, Cartwright Road,
Kingsthorpe,
Northampton, NN2 6HF,
England.
☆ - A, B, C, E, H

KROOKLOK
☎ 0384 374605
Dana Ltd.,
Stour House,
High Street,
Wollaston,
Stourbridge,
W.Midlands. DY8 4PF,
England.
☆ - E

LCB MARKETING
☎ 0322 866313
Greenacres International
Group,
Old Dartford Road,
Farningham,
Kent, DA4 0EB,
England.
☆ - J

LASERLINE CAR ALARMS
☎ 0928 571571
5, Beeston Road,
Stuart Road,
Manor Park,
Runcorn,
Cheshire, WA7 1SG,
England.
☆ - A, C

LIFTSONIC LTD
☎ 0245 422999
PO Box 737,
Chelmsford,
Essex. SCM1 3SD,
England.
☆ - C

LINWOOD ELECTRONIC
☎ 021 358 2171
Austin Way,
Hamstead Industrial Estate,
Birmingham.
B42 1DU,
England.
☆ - A, C

LINX
(See Fanfare Electronics Ltd)

LUCAS
☎ 0827 53344
Unit 7 - 10, Mica Close,

Amington Industrial Estate,
Tamworth,
Staffs, B77 4QH,
England.
☆ - A, C

M.A. DISTRIBUTORS (Maystar)
☎ 0273 720129
Industrial House,
Conway Street,
Hove,
East Sussex. BN3 3LV,
England.
☆ - A, C, D, F, K

MAPLEMAN (UK)
☎ 05394 31700
45 Lake Road,
Ambleside,
Leeds,
LA22 0DF,
England.
☆ - C

META ALARMS
(See Autocar Electrical Ltd)

METRO PRODUCTS
☎ 0883 717644
98 - 102 Station Road East,
Oxted,
Surrey. RH8 0AY,
England.
☆ - E, F

MINDER
See Automanor

MODULAS CAR ALARMS
☎ 0279 870166
1 Pinceybrook,
Takeley,
Herts CM22 6QN,
England.
☆ - A, C

MOSS SECURITY
☎ 0527 584584
Unit 7, Block D, Hemming
Road, Redditch, Worcs. B98
0EA, England.
☆ - A, C

MOTUS SECURITY LOCK
(See Automaxi Ltd)

MUL-T-LOCK
☎ 0536 461111
Welland House,
North Folds Road,
Corby,
Northants,
NN18 9QB,
England.
☆ - E

NATIONWIDE CAR ALARMS
☎ 0923 240525
K & K House, Station Ap-
proach, Rickmansworth
Road, Watford, Herts.
WD1 7LR, England.

☆ - A, N, K

ORVELL ELECTRONICS LTD
☎ 0895 70137
17 Lancaster Road,
Uxbridge,
Middx, UB8 1YW,
England.
☆ - A, C, D

PCA LTD
☎ 0264 334277
Unit 4, Prince Close,
Walworth Industrial Estate,
Andover
SP10 5LL,
England.
☆ - A,C,J

PARK & SECURE
(See EEC Ltd)

PATH GROUP PLC
☎ 021 776 7616
Unit 15, Hayward Industrial
Park,
Thameside Drive,
Castle Bromwich,
Birmingham.
B35 7BT, England.
☆ - J

PATRIOT
(See Hunter Instruments Ltd)

PATROL LINE
(See PCA Ltd)

PHILIPS
☎ 0869 320333
PLS House, Talisman Road,
Bicester, Oxon. OX6 0JX, Eng-
land.
☆ - A, C, F

PHOENIX AUDIO
☎ 021 552 9797
Unit 1B, Demuth Way,
Junction 2 Industrial Estate,
Oldbury,
Warley,
W. Midlands. B69 4LT,
England.

PIRANHA CAR ALARMS
(See Orvell Electronics Ltd)

PORTLAND MARKETING LTD
☎ 0295 250544
Wildmere Road,
Banbury,
Oxon,
OX16 7JU,
England.

QUICKFIT 70 LTD
☎ 0204 62381
Fearnhead Street,
Bolton,
Lancs. BL3 3PE,
England.

QUICKSILVER
☎ 021 772 6426
11-15 Stoney Lane,
Sparkbrook,
Birmingham,
B12 8DL,
England.
☆ - A, C, D

REDSURE
☎ 061 320 0949
Bond Street,
Denton,
Manchester,
M34 3AE,
England.
☆ - I

RETAINACAR LTD
☎ 081 871 1333
72 Bennerley Road,
London, SW11 6DS,
England.
☆ - C, K

RICHBROOK INTERNATIONAL
LIMITED
☎ 071 351 9333
2 Munro Terrace,
Cheyne Walk,
London. SW10 0DL,
England.
☆ - J

RING AUTOMOTIVE
☎ 0532 791791
Geldered Road,
Leeds,
Yorks, LS12 6NB,
England.
☆ - E

ROADPOINT SERVICES LTD
☎ 0952 506522
Roadpoint House,
Dawley,
Telford,
Shrops. TF4 2RJ,
England.
☆ - A

ROADSTAR UK LTD
☎ 0734 321032
Tavistock Industrial Estate,
Ruscombe Lane,
Twyford,
Berks, RG10 9NJ,
England.
☆ - F

SAFEGUARD UK LTD
☎ 0234 355374
Unit 28, College Street,
Kempston,
Beds. MK42 8LU,
England.
☆ - K

SARACEN
(See Quicksilver)

SCORPION VEHICLE SECU-
RITY
☎ 061 777 9666
1 Siemens Road,
Northbank Industrial Estate,
Irlam,
Manchester,
M39 5BL,
England.
☆ - A, C, J

SCOPE MARKETING (COM-
MUNICATIONS UK) LTD
☎ 0803 864569
Unit A, The Scope Complex,
Wills Road,
Totnes Industrial Estate,
South Devon, TQ9 5XN,
England.
☆ - A

SELMAR
☎ 0800 378400
Stella Industries Ltd.,
Springvale Works,
Elland Road,
Brighouse,
W.Yorks. HD6 2RN,
England.
☆ - A, C,

SENTRY SECURITY PRODUCTS
(See Kingavon Ltd.)

SHURLOCK
☎ 0280 822800
Wipac Group Ltd.,
London Road,
Buckingham.
MK18 1BH,
England.
☆ - A

SICURO
(See Nationwide Car Alarms)

SIGMA
☎ 061 905 2023
Unit 1, Manway Business Park,

Canal Road,
Timperley,
Cheshire,
WA14
1TD,
England.
☆ - A, C

SIKURA
(See Demon Products Ltd)

SIMBA SECURITY SYSTEMS LTD
☎ 071 703 0485
Occupation Road,
Walworth,
London SE17 3BE,
England.
☆ - A, C, J, K

SOLA-LARM
(See Solartrack)

SOLARTRACK
☎ 081 595 1218
42 New Road,
Dagenham,
Essex, RM9 6YS,
England.
☆ - A, C

SONAR ELECTRONIC (UK) LTD
☎ 081 671 6818
272A Brixton Hill,
Streatham,
London SW2 1HP,
England.
☆ - A, C

SONIX
(See AAA)

SPARKRITE (STADIUM) LTD
☎ 0922 743676
Lea Woods Road,
Aldridge,
W. Midlands,
England.
☆ - A, C, E

SPYBALL SECURITY SYSTEMS
(See Electrosystems)
SUPALOCK (mail order only)
(See Portland Marketing)

SURVEILLANCE
(See Telesense Security Systems Ltd)

SYKES-PICKAVANT LTD
☎ 0253 721291
Warwick Works,
Kilnhouse Lane,
Lytham St Annes,
Lancs. FY8 3DU,
England.
☆ - J + (Tools and diagnostic
test equipment, multi-meters,
etc).

TELESENSE SECURITY SYSTEMS
LTD.
☎ 0603 765507
2 Bridge Court,
Fishergate,
Norwich,
NR3 1UF,
England.
☆ - B

TEPS WINDOW FILM
(See Tritec Distribution Ltd.)

TERMINATOR
(See Hunter Instruments Ltd)

TRITEC DISTRIBUTION LTD
☎ 0306 628208
Unit 2D, Northland Business
Park,
Warnham,
W. Sussex, RH12 3SH,
England.
☆ - A,C

UNI-VALVE
(See Redsure)

VSL LTD
☎ 0582 423269
Unit 5, Hitchin Road Industrial
Estate,
Oxon Road,
Luton,
Beds LU2 0DZ,
England.
☆ - A, C, K

VAULT X
(See Automobile Imports
Ltd.)

VECTA SECURITY SYSTEMS
(See Liftsonic Ltd.)

WASO LTD
☎ 061 736 0767
Unit 1, Oakwood Estate,
Mode Wheel Road,
Salford,
England.
☆ - A, C

WIDDOWSON-DALEBROOK
ENGINEERS LTD
☎ 0270 661111
Basford Road,
Crewe,
Cheshire CW2 6ES,
England.
☆ - C

ZEMCO LTD
☎ 0203 441428
509 Walsgrave Road,
Coventry,
Warwicks CV2 4AG,
England.
☆ - A, C